Springer Theses

Recognizing Outstanding Ph.D. Research

For further volumes:
http://www.springer.com/series/8790

Aims and Scope

The series "Springer Theses" brings together a selection of the very best Ph.D. theses from around the world and across the physical sciences. Nominated and endorsed by two recognized specialists, each published volume has been selected for its scientific excellence and the high impact of its contents for the pertinent field of research. For greater accessibility to non-specialists, the published versions include an extended introduction, as well as a foreword by the student's supervisor explaining the special relevance of the work for the field. As a whole, the series will provide a valuable resource both for newcomers to the research fields described, and for other scientists seeking detailed background information on special questions. Finally, it provides an accredited documentation of the valuable contributions made by today's younger generation of scientists.

Theses are accepted into the series by invited nomination only and must fulfill all of the following criteria

- They must be written in good English.
- The topic should fall within the confines of Chemistry, Physics, Earth Sciences, Engineering and related interdisciplinary fields such as Materials, Nanoscience, Chemical Engineering, Complex Systems and Biophysics.
- The work reported in the thesis must represent a significant scientific advance.
- If the thesis includes previously published material, permission to reproduce this must be gained from the respective copyright holder.
- They must have been examined and passed during the 12 months prior to nomination.
- Each thesis should include a foreword by the supervisor outlining the significance of its content.
- The theses should have a clearly defined structure including an introduction accessible to scientists not expert in that particular field.

Mercè Pacios Pujadó

Carbon Nanotubes as Platforms for Biosensors with Electrochemical and Electronic Transduction

Doctoral Thesis accepted by
the Universitat Autònoma de Barcelona, Spain

Springer

Author
Dr. Mercè Pacios Pujadó
Universitat Autònoma de Barcelona
Spain

Supervisors
Prof. Dr. Maria José Esplandiu Egido
Centre d'Investigació en Nanociència i Nanotecnologia (CIN2)
Barcelona
Spain

Prof. Dr. Manel del Valle Zafra
Universitat Autònoma de Barcelona
Spain

ISSN 2190-5053 ISSN 2190-5061 (electronic)
ISBN 978-3-642-31420-9 ISBN 978-3-642-31421-6 (eBook)
DOI 10.1007/978-3-642-31421-6
Springer Heidelberg New York Dordrecht London

Library of Congress Control Number: 2012941424

Printed on acid-free paper

Springer is part of Springer Science+Business Media (www.springer.com)

Parts of this thesis have been published in the following journal articles:

M. J. Esplandiu, M. Pacios, E. Bellido, M. del Valle. Carbon nanotubes and electrochemistry. Zeitschrift fuer Physikalische Chemie (Muenchen, Germany). **2007**, 221 (9–10), 1161–1173.

M. Pacios, M. del Valle, J. Bartrolí, M. J. Esplandiu. Electrochemical behaviour of rigid carbon nanotube composite electrodes. Journal of Electroanalytical Chemistry. **2008**, 619–620, 117–124.

M. J. Esplandiu, M. Pacios, L. Cyganek, J. Bartrolí, M. del Valle. Enhancing the electrochemical response of Myoglobin with carbon nanotube electrodes. Nanotechnology. **2009**, 20 (35), 5502.

M. Pacios, M. del Valle, J. Bartrolí, M. J. Esplandiu. Electrocatalyzed O_2 response of myoglobin immobilized on multi-walled carbon nanotube forest electrodes. Journal of Nanoscience and Nanotechnology. **2009**, 9 (10), 6132–6138 (7).

M. Pacios, I. Martín-Fernández, R. Villa, P. Godignon, M. Del Valle, J. Bartrolí and M. J. Esplandiu (**2011**). Carbon Nanotubes as Suitable Electrochemical Platforms for Metalloprotein Sensors and Genosensors, Carbon Nanotubes - Growth and Applications, Dr. Mohammad Naraghi (Ed.), ISBN: 978-953-307-566-2, InTech.

M. Pacios, Nihan Yilmaz, I. Martín-Fernández, R. Villa, P. Godignon, M. Del Valle, J. Bartrolí, M. J. Esplandiu. A simple Approach for DNA detection on Carbon Nanotube Microelectrode Arrays. Sensors & Actuators: B. Chemical. **2012**, 162 (1).

To my parents

A very neat job in a small space

Supervisors' Foreword

The convergence of nano and biotechnology, also known as nanobiotechnology, is a promising up-the-front science topic for improving materials and processes. Within this new field, the development of (bio)sensors from nanomaterials is an exciting area which is already providing real-life applications in the biotechnological, clinical, and environmental fields. In this context, electrochemical nanobiosensors are a preferred choice thanks to their advantages in terms of simplicity, robustness, low cost, miniaturization capability, and integration in microfabricated devices.

Historically, carbon has been widely used as electrode material due to its desirable properties for electrochemical applications. Available in a variety of forms, carbon electrodes are recognized as versatile and easy handling devices, also praised by their rich surface chemistry which has been exploited to influence surface reactivity. More recently, the discovery and popularization of carbon nanotubes (CNTs) has fostered their use as electrode materials, improving reference properties, and propelling in an unprecedented way their electrochemical and electroanalytical applications. CNTs nanometric size and high aspect ratio are the distinct features which have contributed in a larger degree to innovate in electrochemical applications.

The doctoral thesis of Mercè Pacios aims at exploiting the properties of CNTs to design novel electrochemical (bio)sensing devices. Thanks to the prominent electrochemical properties of carbon nanotubes, the design of diverse electrode configurations was possible. This fact, combined with their chemical properties and (bio)functionalization versatility, have made these materials ideal candidates for the development of electrochemical biosensors.

In summary, CNTs have been assayed as electrochemical transducers, finding the factors that most influence the electrode activity. The possibility to arrange CNTs in different geometrical layouts also permitted the design of successively improved transducer platforms and biosensing devices, specially those incorporating enzymes, proteins, and DNA. Exploiting the semiconductor character of CNTs, a last variant has been assayed, which is the field effect transistor configuration (FET). The CNT-FET device, optimized for operating in liquid environment, was used to probe in real-time protein/CNT adsorption and protein/aptamer interactions. Results of this thesis work have shown that these electrochemical and electronic CNT devices can indeed become highly promising for biomolecule sensing and for the sensitive monitoring of biological processes. Throughout this work, new and unexpected gateways have been opened to keep and continue exploring the fascinating world of nanoscience.

Barcelona, May 2012

Dr. Maria José Esplandiu Egido
Dr. Manel del Valle Zafra

Contents

Abbreviations and Symbols

AC	Alternating current
AFM	Atomic force microscopy
BSA	Bovine serum albumin
Cat	Catalase
CC	Chronocoulometry
C-C	Carbon–carbon bond
C_{dl}	Double-layer capacitance
CNT	Carbon nanotube
CNT-FET	Carbon nanotube field-effect transistor
CPE	Constant phase element
CV	Cyclic voltammetry
CVD	Chemical vapour deposition
(D)	Drain electrode
D	Dimensional (1D, 2D, 3D)
DC	Direct current
DMF	Dimethylformamide
DNA	Deoxyribonucleic acid
DOS	Density of electronic states
DPV	Differential pulse voltammetry
ΔE_p	Peak potential difference
EDC	1-ethyl-3(3-dymethyl amino propyl carbodiimide hydrochloride
EDS	Energy-dispersive X-ray spectroscopy
E_f	Fermi energy
EG	Ethylene glycol
EIS	Electrochemical impedance spectroscopy
ESCA	Electron spectroscopy for chemical analysis
ET	Electron transfer
eV	Electron volt
F	Faraday's constant
FET	Field-effect transistor
G	Conductance
(G)	Gate electrode
GC	Glassy carbon

GEC	Graphite-epoxy composite
g_m	Transconductance
GOx	Glucose oxidase
HOPG	Highly ordered pyrolytic graphite
i	Electrical current
K^o	Electron transfer rate constant
LEDs	Light-emitting diodes
LOD	Limit of detection
Mb	Myoglobin
MOSFET	Metal-oxide-semiconductor field-effect transistor
MWNT	Multi-walled carbon nanotube
NHS	N-hydroxysuccinimide
NSB	Non-specific binding
NSL	Nanosphere lithography
1-PBASE	1-(pyrene)butyric acid N-hydroxysuccinimide ester
PBS	Phosphate buffer solution
PECVD	Plasma Enhanced Chemical Vapour Deposition
PEG	Polyetylenglycol
PL	Photoluminescence
PMMA	Polymethylmethacrylate
Q	Charge
R	Ideal gas constant
RBM	Ring breathing modes
R_{ct}	Charge transfer resistance
RIE	Reactive Ion etching
RNA	Ribonucleic acid
ROS	Reactive oxygen species
RTCVD	Rapid-thermal CVD
(S)	Source electrode
SB	Schottky barrier
SB_W	Schottky barrier thickness
SB_H	Schottky barrier height
SCE	Saturated calomel electrode
SELEX	Selection evolution of ligands with exponential enrichment
SEM	Scanning electron microscopy
STM	Scanning tunnelling microscopy
SW	Stone-wales
SWNT	Single-walled carbon nanotube
T	Temperature
TEM	Transmission electron microscopy
TRIS	Tris(hydroxymethyl)aminomethane
UMA	Ultramicroelectrode array
VACNT	Vertically aligned carbon nanotube
V_G	Gate-to-source voltage
V_{SD}	Source-to-drain voltage

V_{thr}	Threshold gate voltage
W	Warburg impedance
XPS	Spectroscopy X-ray photoelectron
Z	Electrochemical impedance

Chapter 1
Introduction

Carbon nanotubes (CNTs) have become one of the most exciting and extensively studied materials of the last two decades. They have captured the interest as nanoscale materials due to their nanometric structure and their impressive list of superlative and outstanding properties. All these ingredients have encouraged their exploitation for promising applications. One of the most interesting ones is related with the use of CNTs as electrochemical platforms for biosensing purposes, the topic in which the present thesis is framed. Accordingly, the main aim of this introductory chapter is to explain the fundamental concepts of the building blocks that constitute this thesis. Therefore, Sect. 1.1 introduces the transducer element: carbon nanotubes (CNT). In this section, the properties of CNTs, their synthesis and purification are explained. Section 1.2 describes the different carbon nanotube platforms developed for biosensor purposes and their fundamentals. Section 1.3 describes the biological recognition elements used for sensing events on the different CNT platforms. Finally, Sect. 1.4 provides an overview of the fundamentals of the main techniques that have allowed characterizing the biosensor devices and following the sensing events.

1.1 Novel Sensing Materials

Nanoscience and Nanotechnology address the study, control, manipulation, and assembly of nanometre(nm) scale components into materials, systems and devices for human interest and needs [1]. The rapid progress of nanotechnology and advanced nanomaterial production offer significant opportunities for designing powerful sensing devices with enhanced performances. Such nanomaterials can exhibit properties and functions different from the ones corresponding to bulk or macroscopic version of them. Additionally, such nanostructures can become suitable materials that favour the integration with biomaterials or biological systems. Under this context, carbon nanotubes have been exploited as a novel material with huge potential in bioanalytical and biosensing applications.

M. Pacios Pujadó, *Carbon Nanotubes as Platforms for Biosensors with Electrochemical and Electronic Transduction*, Springer Theses, DOI: 10.1007/978-3-642-31421-6_1,

1.1.1 Carbon Nanotubes: Structure and Properties of Carbon Nanotubes

This section will be addressed to the discovery, structure and properties of carbon nanotubes. Specifically, special attention will be given to the electronic, mechanical, chemical, electrochemical and optical properties which lead to immediate applications of the CNTs.

1.1.1.1 Discovery, Structure and Electronic Properties

The ability to form very long chains interconnecting C–C covalent bonds allows carbon to form an almost infinite number of compounds. Being one of the most versatile elements, carbon is the chemical basis of all known living systems on Earth; it is the fundamental element of many important biological compounds including sugars, DNA and proteins.

Until few decades ago, it was thought that there were only three structurally different forms or allotropes of carbon since they were abundant in nature: the hardest substance, diamond; one of the softest known substances, the layered graphite and the non-crystalline form, amorphous carbon. However, in the last time we have been witnesses of the discovery of two other carbon allotropes, the fullerene and the carbon nanotubes.

The true identity of the discoverers of carbon nanotubes is a subject of some controversy. For years, scientists assumed that Sumio Iijima, a Japanese physicist, had discovered CNTs in 1991. He published a paper describing his discovery which initiated a flurry of excitement and could be credited by inspiring the many scientists now studying applications of carbon nanotubes. Though Iijima has been given much of the credit for discovering carbon nanotubes, it turns out that the timeline of CNTs goes back much further than 1991.

In 1952, two Russian scientists gave the world its first clear look at carbon nanotubes. LV Radushkevich and VM Lukyanovich published clear images showing multi-walled carbon nanotubes (MWNTs) with a 50 nm diameter (the first known, transmission electron microscope images of carbon nanotubes). Unfortunately, their findings were not given much publicity: their paper was in Russian, published in a Russian journal (Journal of Physical Chemistry of Russia), and was the period of the Cold War.

Before they came to be known as carbon nanotubes, in 1976, Endo, Koyama and Oberlin, observed hollow tubes of rolled up graphite sheets synthesised by a chemical vapour-growth technique [2]. The first specimens observed would later come to be known as single-walled carbon nanotubes (SWNTs). The three scientists were also the first ones to show images of a nanotube with a solitary graphene wall.

In 1981, Russian scientists published more findings. The carbon multi-layer tubular crystals (as they were known then) were made by rolling graphene layers into cylindrical shapes. In 1987, Howard Tennet was given a patent for his cylindrical discrete carbon fibrils.

In 1985 the three Nobel Price Winners, Robert F. Curl, Sir Harold W. Kroto and Richard E. Smalley performed experiments that aimed at understanding the mechanisms by which long chained carbon molecules are formed in interstellar space and circumstellar shells. Graphite was vaporized by laser irradiation, producing a remarkably stable cluster consisting of 60 carbon atoms: the first buckminsterfullerene C_{60} [3].

The research gained new impetus when it was shown in 1990 that C_{60} could be produced in a simple arc-evaporation apparatus readily available in all laboratories. It was just by analysing samples from such evaporator that the Japanese scientist Sumio Iijima of the NEC Corporation discovered fullerene-related carbon nanotubes in 1991 [4]. These were elongated fullerenes with diameters as small as 0.7 nm and lengths of up to several microns which were termed carbon nanotubes.

The graphene layers have become the starting point to explain the structure of carbon nanotubes. A single-walled carbon nanotube is a rolled-up tubular shell of graphene sheet which is made up of benzene-type hexagonal rings of carbon atoms. The structure is conveniently expressed in terms of a one-dimensional unit cell. Indeed, the appearance of a closed cage in CNTs can be easily rationalized by considering the presence of high energy dangling bonds at the boundaries of a finite graphene layer. The total energy of carbon atoms in a layer can be reduced by promoting the formation of a closed structure which eliminates the dangling bonds, even at the expense of increasing the strain energy.

The way the graphene sheet is wrapped up can be described by a pair of indices (*n*, *m*) that define the chiral vector, $\vec{C} = n\vec{a}_1 + m\vec{a}_2$, *in* which $\vec{a}_1$ and $\vec{a}_2$ are the basis vectors of the hexagonal graphene lattice [5–10] as shown in Fig. 1.1. Three different types of nanotube structures can be generated by rolling up the graphene layer: zigzag ($m = 0$), armchair ($n = m$) and chiral nanotubes (the rest of vectors).

Although CNTs are closely related to 2D graphene, the tube curvature and the quantum confinement[1] in the circumferential direction of the nanotube bring about the unique properties that make CNTs different from graphene. One of these unusual properties is the electronic conductivity which strongly depends on the chirality and CNT diameter. CNTs can exhibit singular electronic band structures and can show metallic and semiconducting behaviour. As a general rule, (*n*, *m*) tubes with $n - m$ being an integer multiple of 3 are metallic while the remaining tubes are semiconducting. The band gap of semiconducting tubes can be approximated by the relation $E_g = 0.8\,\mathrm{eV}/d$, with *d* being the diameter of the nanotube [11, 12]. Therefore, the bigger the diameter, the more metallic behaviour is found.

The verification of the electronic properties of carbon nanotubes was of great interest in light of the theoretical predictions. It turned out to be very challenging to measure them due to the small diameter of the tubes. At the beginning, many of

[1] The electron wavelength around the circumference of a nanotube is quantized due to periodic boundary conditions. Along the tube the electrons are not confined.

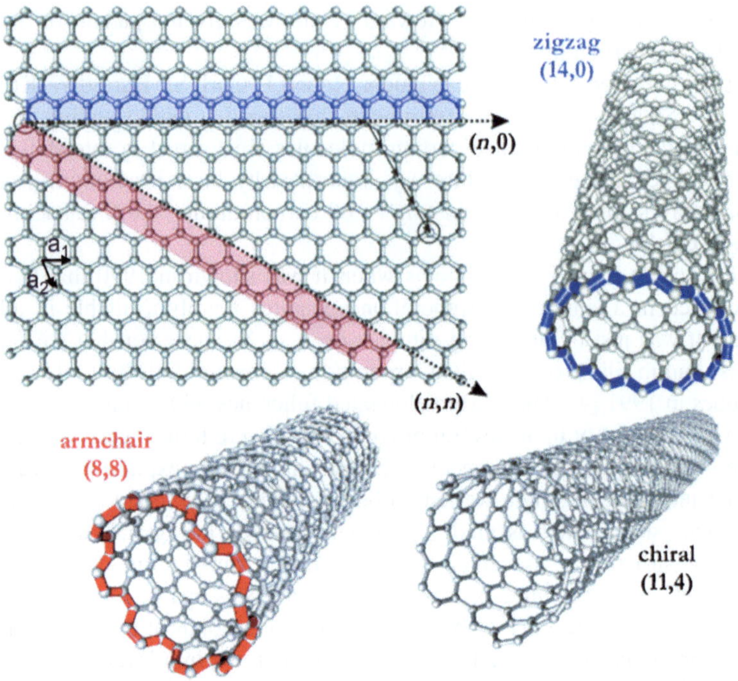

Fig. 1.1 Roll-up of a graphene sheet leading to three different types of SWNTs. Reprinted with permission from Ref. [10]

the studies on the electrons properties were on bulk CNT material by performing, for instance, electron spin resonance. Then, a more sophisticated methodology was employed based on scanning tunnelling microscopy (STM) which allowed addressing individual carbon nanotubes. The tip of the STM was used as a spectroscopic probe to collect the tunnelling conductance of each tube, providing a direct measure of the local electronic density of states (DOS)[2] of the nanotube. Since the STM has the additional power to obtain atomically resolved images of the tube's hexagon lattice, the electronic structure could be correlated with the chiral structure of the tube and with the carbon nanotube semiconducting or metallic properties.

Figure 1.2 shows the density of electronic states for a metallic and a semiconducting carbon nanotube. The DOS is not a continuous function of energy as in the case of bulk 3D materials (e.g. graphite) but they present discontinuous spikes which are typically present in one-dimensional materials (Van Hove singularities).

---

[2] Density of states (DOS) of a system describes the number of electron states per unit volume and per unit energy that are available to be occupied.

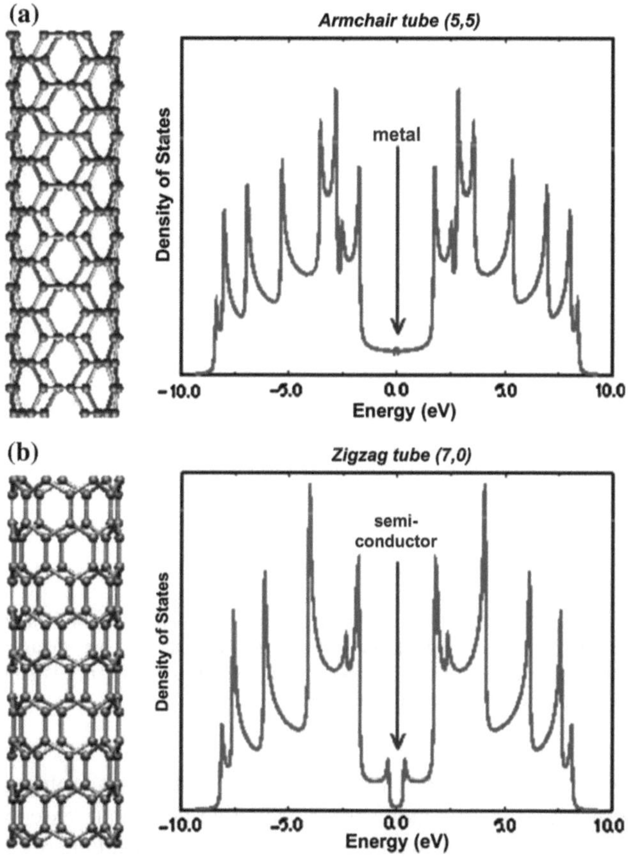

Fig. 1.2 Density of states (DOS) for a metallic (**a**) and semiconducting (**b**) carbon nanotube with the sharp Van Hove singularities. The density of states at the Fermi energy (E = 0) (Fermi energy is the energy of the highest occupied state at the absolute zero temperature.) for the metallic tube is finite but zero for the semiconducting one, providing a band-gap in the latter case. The DOS at the right of the Fermi energy are part of the conduction band (unoccupied states) whereas the one at the left corresponds to the valence band (occupied states)

Another way to verify the electronic properties of carbon nanotubes is by performing electron transport studies by contacting them in a field-effect transistor (FET) configuration. In this configuration, three electrodes are employed. Two of them (source and drain electrodes) are used to contact the nanotube and allow the flow of current through the tube when a voltage is applied between them. The third electrode (gate) is separated from the nanotube by a dielectric, and when a voltage is applied through such electrode, a modulation of the tube current is produced. The charge carriers of the carbon nanotube are modulated in a big extent if the contacted nanotube is a semiconducting but the nanotube conductance remains almost constant if the nanotube is metallic. Therefore, by sweeping the

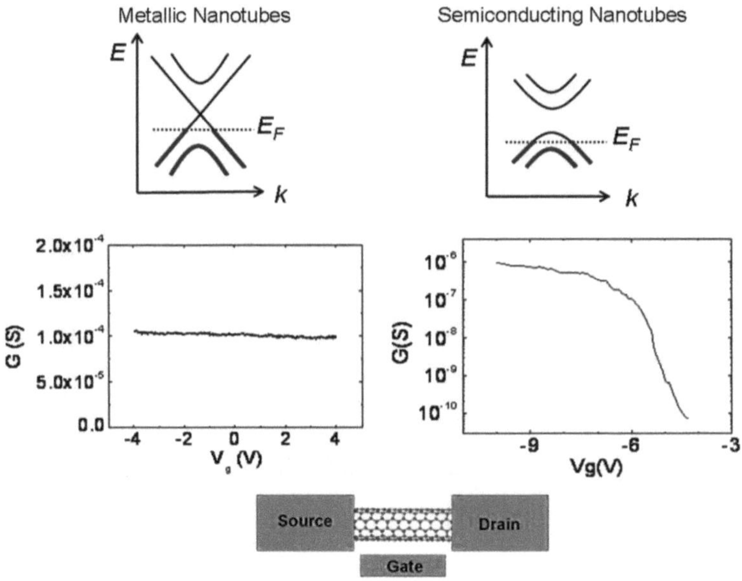

Fig. 1.3 Energy bands for a metallic and a semiconducting carbon nanotube. Note that in the case of the metallic tube, the conductance (G) as a function of the gate voltage (V_G) remains almost constant but in the case of the semiconducting one, the conductance is highly modulated by the gate voltage due to the presence of the band-gap

gate voltage of a CNT-FET and collecting the conductance profile one can discriminate between a metallic and a semiconducting behaviour. Sweeping the gate voltage shifts the Fermi level which allows probing the band energy of the tubes. In the case of the metallic CNTs (Fig. 1.3), there are always electronic states for the conduction which makes the conductance profile to keep unchangeable as the gate voltage is varied. However, in the case of the semiconducting tubes, the presence of the band gap makes the current change several orders of magnitude from a certain value (state "on") up to zero when the gate voltage is probing the region of the gap (state "off").

Moreover, carbon nanotubes can exhibit other distinct electronic characteristics such as a ballistic flow of electrons with long electron mean-free paths and the capability of supporting very large current densities [13–15]. For instance, carbon nanotubes have been shown to withstand current densities up to 10^9 A cm^{-2}, a density which is about 2–3 orders of magnitude larger than in metals such as Al or Cu.

The bonding in CNT is basically sp^2; however, the circular curvature, apart from yielding quantum confinement and quantized conductance, also causes σ–π rehybridization, an effect that is stronger as the CNT diameter decreases [16–18]. In such case, three σ bonds are slightly out of plane and for compensation the π orbitals are more delocalized outside the tube. That is a big difference with respect to graphite in which the sp^2 hybrid orbital form three in-plane σ bonds with an out-of-plane π orbital. The electron cloud distortion induced by the curvature,

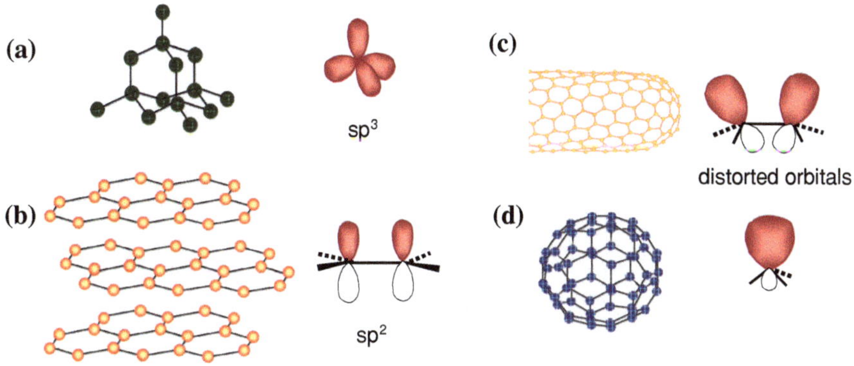

Fig. 1.4 Bonding structures for different carbon materials: **a** diamond, **b** graphite, **c** carbon nanotubes and **d** fullerenes

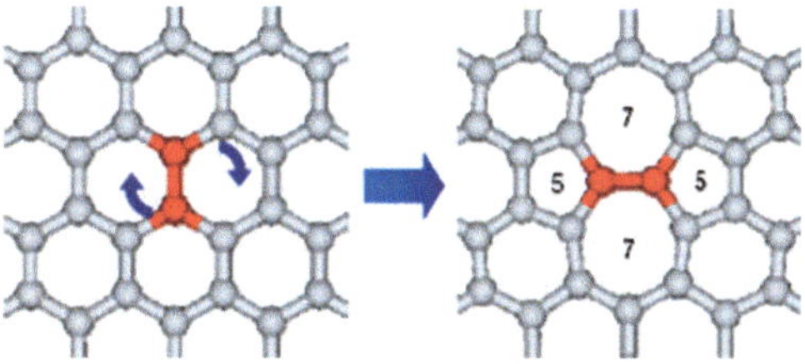

Fig. 1.5 Atomic arrangement of the Stone–Wales (SW) model. SW transformation leading to the 5–7–7–5 defect generated by rotating a C–C bond in a hexagonal network. Reprinted with permission from Ref. [25]

which yields a rich π-electron conjugation outside the tube, can make CNT more electrochemically active and electrically and thermally more conductive (Fig. 1.4).

The latter fact together with the combination of size (small diameter), structure and topology confer nanotubes remarkable mechanical properties such as high stability, strength and stiffness together with low density and elastic deformability.

However, the mentioned mechanical properties can be affected by imperfections in their structure. That is, carbon nanotubes (CNTs) can have a lot of defects such as vacancies, metastable atoms, pentagons, heptagons, Stone–Wales (SW or a pair of 5–7 rings) defects [19], discontinuities of walls, and heterogeneous atoms [20]. Therefore, being SW defects stable and commonly present in CNTs, they are believed to play key roles not only in the mechanical [21] properties of CNTs, but also in electronic [22], and chemical [23] properties. Figure 1.5 shows the pairs of 5–7 (pentagon–heptagon) defects. They are created at high strains in the nanotube lattice. This leads a step-wise diameter reduction (localized necking) of the nanotube [24].

Thus, the σ–π rehybridization, the high surface area and the presence of potential topological defects constitute key points for chemical, electrochemical and

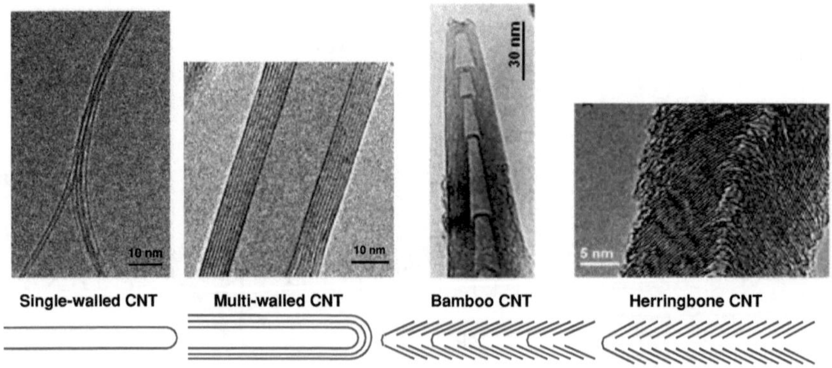

Fig. 1.6 Different structures of Carbon Nanotubes. From *left* to *right*, Single-walled CNT, Multi-walled CNT, Bamboo CNT and Herringbone CNT. Reprinted with permission from Ref. [33]

biochemical reactivity. As a result, molecular adsorption, doping and charge transfer can be more facilitated if compared with its analogue graphite.

So far we have considered CNTs formed by a single graphene wall (SWNT) with diameters ranging generally from 0.4 up to 1.5 nm and with an important diversity in electronic conductivity, from semiconducting to metallic behaviour. However, CNTs can also be made by multiple graphene layers disposed in different arrangements. Such structures, known as multi-walled carbon nanotubes (MWNTs), have typically diameters in the range of 2–100 nm and metallic behaviour.

MWNTs can be described as an assembly of concentric single-walled carbon nanotubes with different diameters and chiralities (hollow MWNTs) or with the graphene planes forming an angle with respect to the axis of the tube (herringbone or bamboo MWNTs), see schemes in Fig. 1.6. The bamboo-like tubes differ from the herringbone ones in that some of the graphene layers are periodically closed along the length of the tube forming compartments. The main difference between the hollow morphology and the herringbone and bamboo ones lies on the high density of terminating edge planes that contain the last two morphological variations of MWNTs.

1.1.1.2 Mechanical Properties

As stated before, the sp^2 bond between the individual carbon atoms makes carbon nanotubes extremely stiff and resistant to physical forces. The Young's modulus[3] of perfect CNTs has been theoretically calculated using different models [26–28].

[3] Young's modulus is a measure of the stiffness of an elastic material and is a quantity used to characterize materials.

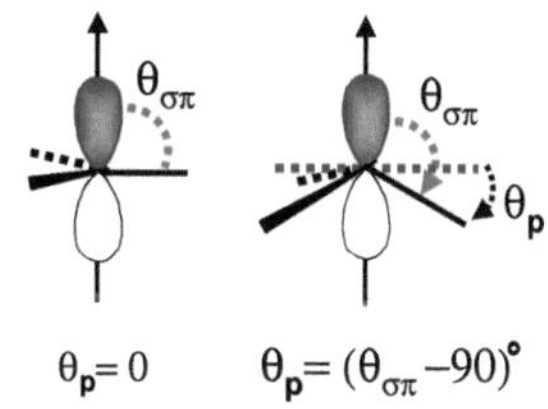

Fig. 1.7 Scheme of the pyramidalization angle in distorted sp^2 bonding in comparison with a trigonal structure

Commercially available SWNTs can achieve Young's modulus as high as 1,000 GPa. The precise value depends on the diameter (chirality) and the amount of disorder in the nanotube walls.

The maximum tensile strength[4] of SWNTs has been reported to be 30 GPa [29]. The same authors found a tensile strength of 63 GPa for MWNTs [30], while Demczyk et al. estimated the tensile strength of MWNTs to be 150 GPa [31]. Due to the nonlinearity of the radial compression, moduli from 9.7 to 80.0 GPa where found when com pressing a 10 nm tube from 26 to 46 % [32].

1.1.1.3 Chemical Properties

The high specific surface of CNTs together with the σ–π rehybridization and the presence of structural defects, facilitate different chemical processes such as (bio) chemical derivatization, intercalation, molecular adsorption, doping, charge transfer, etc. [15–17, 26–28, 33]. These characteristics are being considerably exploited for different applications such as biochemical and chemical sensing, energy storage or separation techniques.

Although graphite, fullerene and carbon nanotubes are built from the same basic element, C, the chemical reactivity is substantially different among them. The chemical reactivity is higher for a CNT than for a graphene layer but lower if compared with fullerene [16–18, 34]. Such behaviour can be closely related to the surface curvature of the carbon structure [15–17, 26, 27, 33]. The reactivity of fullerenes is mainly due to the high strain generated by the spherical geometry which affects the pyramidalization angle θp among carbon bonds which is defined as the difference between the σ–π orbital angle and 90° [34]. θp is a quantitative measure of the curvature at a tri-coordinated carbon atom. As illustrated in Fig. 1.7, this angle quantifies the deviation of a carbon atom from the plane passing through the three adjacent carbon atoms. The higher the θp, the more reactive the specie is. In a pure sp^2 hybridization (trigonal, graphene), $\theta p = 0$. Fullerenes have a θp of 11.6° due to 2D curved structure which results in sp^2–sp^3 or σ–π mixing, as discussed above with CNT (Fig. 1.4). In this case, reactions that lead to a strain relief are enhanced such as the ones related to the addition chemistry [16–18, 34–36].

[4] The tensile strength of a material is the maximum amount of tension that it can take before failure, for example breaking.

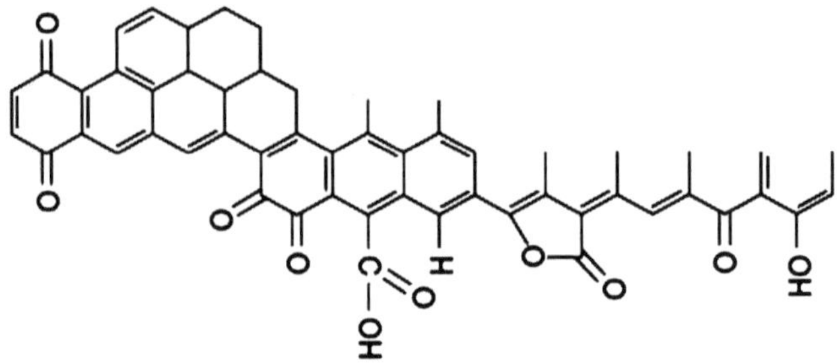

Fig. 1.8 Different oxygenated functionalities that can appear at CNT edges

In the case of CNTs, the strain appears as a consequence not only of the curvature-induced pyramidalization but also from π orbital misalignment between adjacent pairs of conjugated carbon atoms which favours the chemically reactivity of the CNT walls as compared to that of flat graphene layers [34]. Furthermore, since pyramidalization angles and π-orbital misalignment angles of CNT scale inversely with the diameter of the tubes, a differentiation is expected between the reactivity of carbon nanotubes of different diameters. A higher surface reactivity with decreasing diameter is thus expected [35].

The reactivity of different graphitic carbon allotropes is also characterized by having chemical anisotropy. Taking as a reference, for instance, the layered structure of Highly Ordered Pyrolytic Graphite (HOPG), it is known that its basal plane is more inert than the highly reactive edges which contain unsatisfied valences or dangling bonds prone to reaction with oxygen or water [37]. Such anisotropy is also observed in adsorption processes. The same anisotropy in the chemical reactivity can be considered on CNT, since the walls behave differently than the ends which can be either capped or just finishing in terminal edges [9, 26, 30, 31]. The chemical reactivity is increased at the capped ends (due to the presence, for instance, of pentagons) or at the edges of an open nanotube, which become the more reactive sites of the CNT irrespective of its diameter. Such anisotropy in the chemical reactivity will also have strong implications from the point of view of electrochemistry and will be approached in the next sections.

Although the CNT ends exhibit a higher chemical reactivity than the walls, the presence of local defects on the walls constitutes also efficient reactive sites. For instance, when CNTs are subjected to oxidative acid treatment, the local generation of carboxylic, ketones, alcohols and ester groups leads to active sites that can be profited for (bio)molecular anchoring [26, 30–32, 34]. Such acid treatments also open the CNT capped ends, bringing about an increase of oxygen rich terminal groups.

Figure 1.8 depicts a general overview of the different oxygen functionalities that can appear at the edges of CNTs [17, 37]. Such defects together with the reactivity associated with the generated strain from the curvature and π misalignment are very important for the attachment of a wide diversity of molecules. In this context, it is expected that the chemical reactivity is increased in MWNT with higher density of reactive edges, being the favoured structures the bamboo or herringbone CNTs.

1.1.1.4 Electrochemical Properties

In general, the electrochemical performance of carbon materials is basically determined by the electronic properties, and given its interfacial character, by the surface structure and surface chemistry (e.g. surface terminal functional groups or adsorption processes) [37, 38]. Such features will affect the electrode kinetics, potential limits, background currents and the interaction with molecules in solution [37]. From the point of view of electroanalysis, the remarkable benefits of CNT-modified electrodes have been widely praised, including low detection limits, increased sensitivity and decreased overpotentials [39–42]. Thus, the enhancement of the CNT electrode kinetics to different redox species, in terms of increased faradaic current and minimization of the redox overpotential, has made the nanotubes to be considered as electrocatalytical materials, a topic that will be discussed later on.

The electronic properties of CNTs, and specially their band structure, in terms of density of electronic states (DOS), are very important for the interfacial electron transfer between a redox system in solution and the carbon electrode. There should be a correlation between the density of electronic states and electron transfer reactivity. As expected, the electron transfer kinetics is faster when there is a high density of electronic states with energy values in the range of donor and acceptor levels in the redox system [37]. Conventional metals (Pt, Au, etc.) have a large DOS in the electrochemical potential window, whereas carbon electrodes have a considerable lower DOS (especially in the case of HOPG) that can be incremented by the introduction of disorder. For instance, the low DOS of HOPG can be reflected in a remarkable low double layer capacitance when the basal plane is considered [37, 38]. Interestingly, SWNTs have additionally a non-trivial variety of diameter-dependent DOS distribution and as a consequence, a very rich electrochemical behaviour is expected. SWNTs of different diameters will yield different rate constants due to the structure-dependent variations in the DOS which could be tracked by electrochemistry [43].

As mentioned, the chemical anisotropy characteristic of graphitic materials is also present in their electronic properties. For instance, it is well known that in the layered structure of HOPG, the electronic conductivity measured in the direction perpendicular to the basal plane, is about four orders of magnitude lower than that measured in the edge plane [38]. Such anisotropy also influences the electrochemical behaviour inducing an electrochemical anisotropy as well. It has been demonstrated that in graphite, the basal plane exhibits slow electron transfer kinetics whereas the reactive edge sites increase the electron transfer rate. For instance, the electron transfer rate constant, k^o, for a $Fe(CN)_6{}^{3-/4-}$ redox system is smaller than 10^{-6} cm s^{-1} in the case of basal HOPG and around 0.06–0.10 cm s^{-1} in the case of edge HOPG [38]. The same electrochemical anisotropy has been attributed to CNTs. The open ends of carbon nanotubes have been likened to the edge planes of HOPG whereas the tube walls are suggested to have similar electrochemical properties to those of basal HOPG [44–48].

Figure 1.9b shows the typical electrochemical response that was found when comparing CNT modified electrodes, edge plane and basal plane pyrolytic

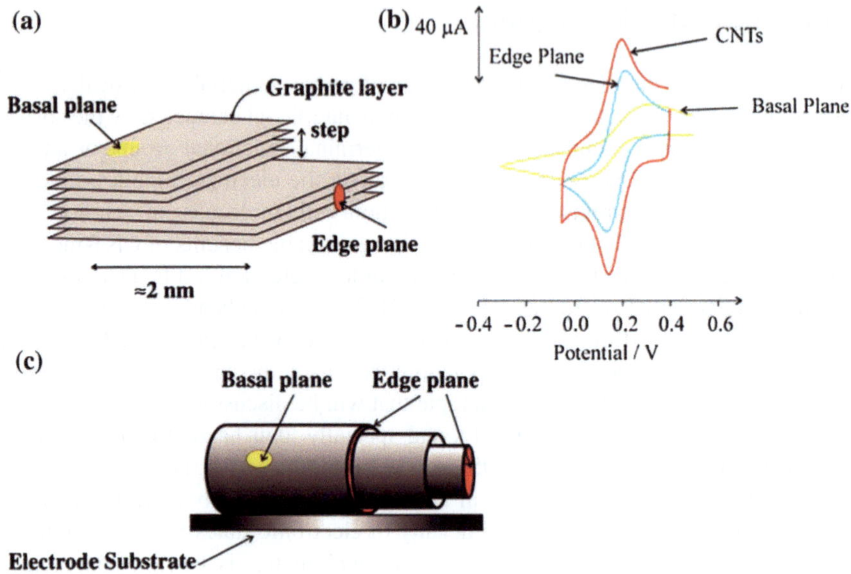

Fig. 1.9 **a** Schematic representation of the layered structure of HOPG with the indication of the different edge and basal planes; **b** voltammetric response of $Fe(CN)_6^{3-/4-}$ using basal plane, edge plane pyrolytic graphite and CNT electrodes; **c** scheme of a MWNT on an electrode surface showing the edge plane and basal plane like sites. Reprinted with permission from Ref. [50]

graphite electrodes. Basal pyrolytic graphite exhibits higher peak separations (which depends on the surface edge defect density) and consequently lower electron transfer rates [47]. However, edge plane pyrolytic graphite electrodes and carbon nanotube modified substrates show smaller and similar peak separation corresponding to fast electron transfer kinetics. Such finding only demystified the claim that CNTs provide novel electrochemical attributes such as those electrocatalytical effects [46–49]. Indeed, CNT can exhibit electrocatalytic effects but in the same way as can be found on edge pyrolytic graphite systems. That only reinforces the idea that the edge-plane like sites (like the ones at the open ends of nanotubes) are the ones responsible for such electrocatalytic behaviour. However, the electrochemical anisotropy on CNT should not also lead us to completely disregard the intrinsic electrochemical properties of the CNT walls.

One distinct advantage of the CNTs comes from the nanometre size which can also be manipulated to design novel electrode architectures [49, 51] or for signal amplification purposes [36, 37, 48]. Related with the latter issue, CNTs-systems exhibit a high effective surface area which is beneficial not only for enhancing electrochemical currents of diffusing electroactive species but also for allowing the load of high density electrochemical active (bio)molecules which can improve electrochemical signals [52, 53].

It is important to point out that not all redox systems will exhibit electrocatalytic activity when probed on high density edge CNT electrodes or edge pyrolytic graphite. Such phenomenon depends on the particular mechanism of the redox system [37]. It is important to mention that among the redox species, there are some of them whose electrochemical behaviour only depend on the carbon DOS but are insensitive to the surface structure/morphology or chemistry [37, 38]. That is the case of many outer-sphere[5] redox species like $Ru(NH_3)_6^{2+/3+}$, ferrocene and anthracene. However, electrocatalytic effects are relevant on redox systems which involve specific chemical interactions with the electrodes and therefore are very dependent on the surface structure and surface oxygen functionalities [37, 38]. In general, an electrocatalytical behaviour is observed, correlated with redox species that are very sensitive to the surface characteristics such as the inner-sphere[5] redox species NADH, O_2, catecholamines and $Fe(CN)_6^{3-/4-}$ [54, 55]. Moreover, at this point it is important to recall the relevance of surface oxygen functionalities on electrocatalysis processes. Edge defect sites or unsatisfied valence are prone to react with oxygen and water. Such moieties can play an important role in electrocatalytical processes by enhancing the electron transfer kinetics.

Another relevant factor affecting the electrochemical performance of carbon nanotubes is adsorption. CNTs exhibit high specific surface area which can develop high polarizability and additionally oxygen functionalities which can contribute with permanent dipoles [37]. Both effects can induce dipole interactions with adsorbates and favour the adsorption, which can be assisted by other interactions such as hydrophobic ones. Adsorbates can strongly alter the electron transfer rates and the electrocatalytical activity. Positive effects can be found with the adsorption of some metalloproteins on CNTs which facilitates the electron exchange between the redox core and the carbon electrode and allows electrocatalytic activity to be monitored. Negative aspects are found when impurities are adsorbed on CNT substrates decreasing dramatically the electrochemical performance of the electrodes. In that case, activation procedures have been developed for removing such impurities. Precisely, surface preparation, as briefly discussed later on, is very important for the electrochemical performance on CNT electrodes.

To summarize, one can say that the electrochemical performance of CNT electrodes is correlated to the DOS of the CNT electrode with energies close to the redox formal potential of the solution species. The electron transfer and adsorption reactivity of CNT electrodes are remarkably dependent on the density of edge sites/defects which are the more reactive sites for that processes, increasing considerably the electron transfer rate. However, it is also worth mentioning that an exaggerated increase of edge sites/defects can also become detrimental for the CNT electrochemical properties due to the disruption of their electronic

---

[5] Outer-sphere or non-bonded electron transfer refers to an electron transfer (ET) event that occurs between chemical moieties that remain separate species before, during, and after the ET even, with electron tunnelling from one to the other, probably across the salvation layer. This is opposed to inner sphere electron transfer in which the electroactive species interact strongly with the surface. The ET undergoes by a chemical bridge being the reactants, intermediates or products strongly adsorbed on the electrode surface.

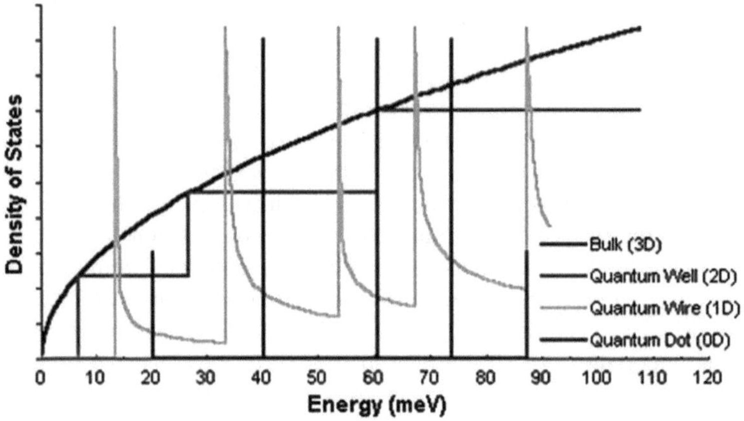

Fig. 1.10 A bulk 3D material has continuous DOS, but a 1D wire has Van Hove singularities

conductive properties. Additionally, surface oxygen functionalities can exert a big influence on the electrode kinetics. However, not all redox systems respond in the same way to the surface characteristics or can have electrocatalytical activity. That is very dependent on their redox mechanism. Moreover, the nanometre size and the high surface area are the key factors in the electrochemical performance of the carbon nanotubes.

1.1.1.5 Optical Properties

Whereas mechanical, chemical and electrochemical properties of the carbon nanotubes are well established and have immediate applications, the practical use of their optical properties is yet unclear. In particular, CNTs seem to be promising candidates for nanoscale light-emitting diodes (LEDs) [56, 57], photo-detectors [58] based on a single nanotube, bolometer [59] and optoelectronic memory [60] devices.

The remarkable CNT capabilities for light detection applications are closely related to their quasi one dimensional structure. As mentioned, the typical feature of one-dimensional crystals is that their DOS is not a continuous function of energy, but it descends gradually and then increases in a discontinuous spike. In contrast, three-dimensional materials have continuous DOS. The sharp peaks found in one-dimensional materials are the discussed before Van Hove singularities and are shown in Fig. 1.10.

Van Hove singularities lead to optical transitions which are sensitive to the chirality of the tubes. Optical transitions occur between the v_1–c_1, v_2–c_2 and can be labelled as E11, E22, etc. (Fig. 1.11). Crossover transitions c_1–v_2, c_2–v_1, etc., are dipole-forbidden and thus are extremely weak, but they can be possibly observed using cross-polarized optical geometry [61]. As the energies between the Van

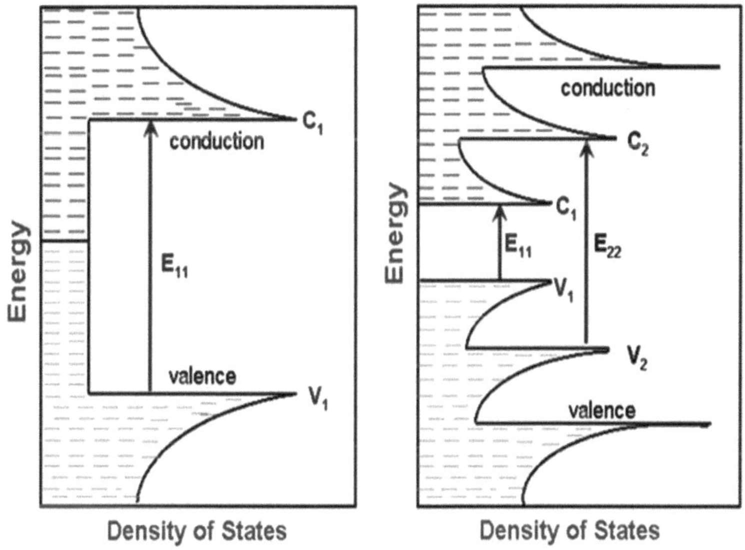

Fig. 1.11 Typical densities of states as a function of energy for metallic (*left*) and semiconducting (*right*) single-wall carbon nanotubes. $v_1 \rightarrow c_1$ corresponds to the 'first Van Hove' optical transition. $v_2 \rightarrow c_2$ corresponds to the 'second Van Hove' optical transition

Hove singularities depend on the nanotube structure, by varying this structure, one can tune the optoelectronic properties of carbon nanotube. Such fine tuning has been experimentally demonstrated using UV illumination of polymer-dispersed CNTs [62].

Optical transitions are rather sharp and strong. Consequently, it is relatively easy to selectively excite nanotubes having certain (*n*, *m*) indices, as well as to detect optical signals from individual nanotubes.

Experimentally, optical properties of CNTs are studied by optical ellipsometry, electron energy loss spectroscopy, reflectivity measurements, absorption experiments, photoluminescence, Raman spectroscopy, etc.

Among these techniques, photoluminescence (PL) is widely used to deduce (*n*, *m*) indices. In PL an electron in a nanotube absorbs excitation light via E_{22} transition, creating an electron–hole pair (exciton). Both electron and hole rapidly relax (via phonon-assisted processes) from c_2 to c_1 and from v_2 to v_1 states, respectively. Then they recombine through a c_1–v_1 transition resulting in light emission. As no excitonic luminescence can be produced in metallic tubes, this permits to identify semiconducting nanotubes.

Besides, carbon nanotubes have quite interesting Raman spectra due to resonance effects and electron–phonon interactions [63–66]. There is very strong excitation wavelength dependence of the spectra resulting from the CNT electronic band structure. Features in the Raman spectra, which will be discussed in detail later on, are diagnostic of the CNT structural quality.

1.1.2 Synthesis of CNTs

Carbon nanotubes can have different individual structures, morphologies and properties, as well as different collective arrangements and emerging properties, all of which are determined by the method of preparation and further processing [67]. Hence, a wide variety of synthetic methods have been developed to produce the desired materials and properties for specific scientific studies or technological applications.

Typical methods for carbon nanotubes synthesis are electric arc discharge [68], laser ablation [69] and chemical vapour deposition (CVD) [70]. A scheme of the three methods is depicted in Fig. 1.12.

The early processes used for CNT production were arc discharge and laser ablation approach; they allowed synthesis of SWNTs in relatively large (gram) amounts [71]. Arc-discharge and laser ablation rely on the condensation of carbon atoms generated from evaporation of solid carbon sources. The arc method creates CNTs through arc-vaporization of two carbon (graphite) electrodes placed end to end, separated by approximately 1 mm, in an enclosure that is usually filled with inert gas at low pressure. A direct current of 50–100 A, driven by a potential difference of approximately 20 V, creates a high temperature discharge between the two electrodes which vaporizes the surface of one of the carbon electrodes and forms a small rod-shaped deposit on the other electrode. MWNTs produced by arc discharge are long and straight tubes closed at both ends with graphitic walls running parallel to the tube axis. For the growth of single-walled tubes, a metal catalyst (such as cobalt, iron or nickel) mixed with graphite powder is needed. Growth temperature in the arc discharge is considerably higher than other CNT production methods (typically in the range of 2,000–3,500 °C). Consequently, crystallinity and structural perfection of arc-produced CNTs are higher (due to mechanisms of annealing of defects at high temperatures) and a significantly higher yield per unit time is obtained in comparison to other methods. In the laser ablation method, the synthesis is carried out in a horizontal flow tube under a flow of inert gas at controlled pressure. In this set-up, the flow tube is heated to ~1,200 °C by a tube furnace. Laser pulses enter the tube and strike a target consisting of a mixture of graphite and a metal catalyst such as Co or Ni. SWNTs condense from the laser vaporization plume and are deposited on a collector outside the furnace zone.

Both methods provide high quality CNTs with very little defects. Nevertheless, the equipment requirements and the large amount of energy consumed make them less favourable for nanotube production. Moreover, they produce highly entangled CNT bundles which are difficult to manipulate and assemble for building addressable structures.

The third method, based on CVD, has been widely used to grow CNTs in recent years and performed in our laboratory to understand different issues of this thesis work. In this case, the formation of tubes occurs as a result of thermal decomposition of a liquid or gas phase carbon source such as CO or hydrocarbon (C_2H_2 CH_4, etc.), catalyzed by metallic particles (typically Fe for SWNT, Co or Ni

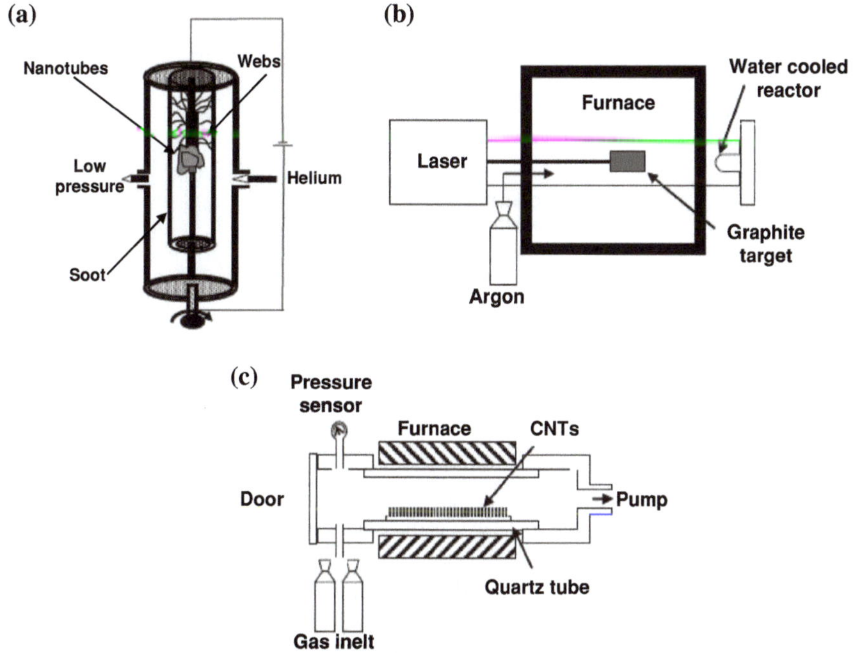

Fig. 1.12 Schematics of the three established CNT production techniques: **a** electric arc discharge, **b** laser ablation and **c** chemical vapour deposition (CVD)

nanoparticles for MWNT), which also serve as nucleation sites for the initiation of carbon-nanotube growth. It operates at considerable lower temperatures (typically between 600–1,200 °C; the lower temperatures for MWNT) with the handicap of bringing about more defective CNTs than in the case of the other methodologies. The process is sensitive to the catalyst structuring, as well as to the reaction conditions. The growing interest in the CVD process lies on the greater possibilities of this method for successful synthesis of well-separated individual SWNTs and also aligned SWNTs or MWNTs with more controllable diameter and length. It has also the advantage of being the only technique that allows synthesis of CNTs directly on a substrate or wafer, thus facilitating device integration (these nanotubes can be directly used to fabricate nanoscale electronics).

The entire CVD process mainly consists of two steps, namely catalyst preparation and CNT growth.

The catalyst is generally prepared by dispersing a solution of a metal cation on a substrate. For instance, different concentrations of iron ($Fe(NO_3)_3$) in isopropanol or of a catalyst mixture ($Fe(NO_3)_3/MoO_2(AcAc)_2$/alumina) nanoparticles in methanol, are deposited over portions of silicon wafers with silicon oxide thermally grown via spin-coating. In the next phase, the sample is introduced in the quartz tube of the CVD reactor with a flow of argon (the catalyst has to be in an oxygen-free atmosphere); then, the carbon source enters to produce the growth of the nanotubes.

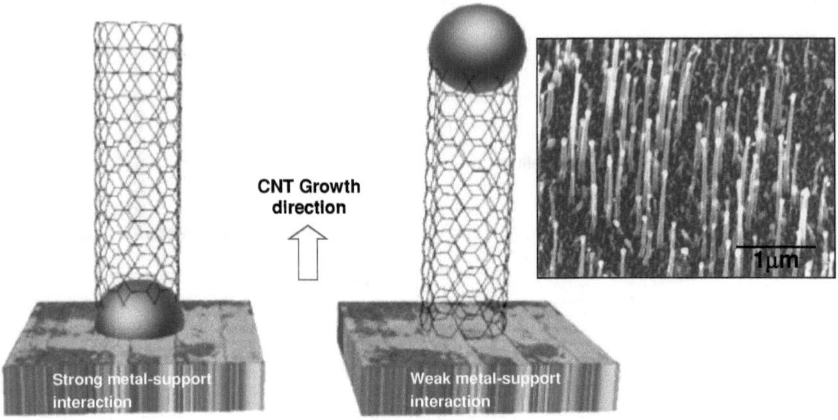

Fig. 1.13 Carbon nantube mechanistic models: base-growth mechanism (*left*) and tip-growth (*right*) with a SEM image showing Ni catalyst particle at the top of the CNT

One model that can explain the mechanism of CVD growth is the "yarmulke" one [72]: the carbon atoms released by decomposition of hydrocarbons will form graphitic sheets covering the catalytic metal surface and in the nano-sized particle, an hemispherical cap covering the exposed portion of particle will be generated. Under the right conditions, this carbon cap will peel off from the metal nano-particles as long as carbon atoms are added to the edges of the lid. Consequently, the structure takes the form of a tube whose edge is in contact with the nanoparticle and where carbon atoms continue being added. In this mechanism, the diameter of the nanotube is correlated with the catalyst particle diameter. To lift-off the lid and begin the growth of CNTs, the metal particle must have sufficient curvature (for example, must be small enough) so that graphite layers of the cover could be sufficiently tight and the lid could be lifted and the addition of carbon in tubular form could be energetically favourable. In other words, the nucleation and growth will happen only if the metal catalyst particle does not exceed a certain size. If the particle is too large and the graphite layer is not sufficiently tight and does not lift off, the particle simply is covered with carbon.

Taking as an example methane as precursor gas, the catalytic effect of the metal breaks down the methane into carbon (graphitic and amorphous carbon competing) and hydrogen; this reaction is enhanced in high temperature conditions: $CH_4 \leftrightarrows C\ (C_{graphite} + C_{amorphous}) + 2H_2$. The catalyst in the form of nanoparticles also serves as growth centre to form the basis of nanotubes [73]. Therefore, the diameters of nanotubes are correlated with the diameters of the particles, thus leading to control its diameter at its creation [74].

During the CVD growth processes, the catalyst nanoparticle can exhibit two different behaviours depending on the interaction substrate/particle. When the catalyst remains anchored to the substrate (strong interaction), the synthesis mode is

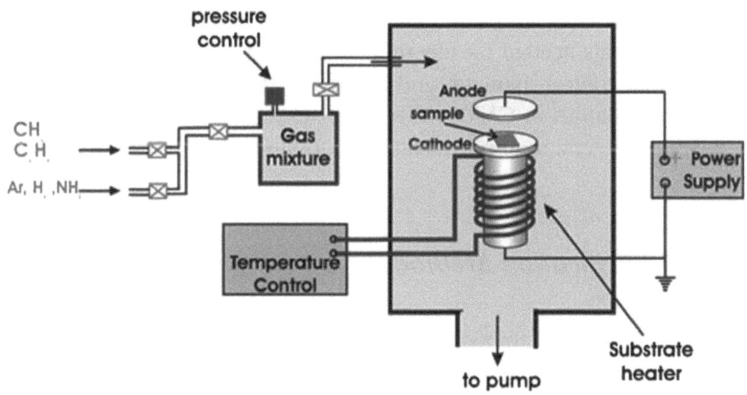

Fig. 1.14 Plasma-enhanced CVD reactor

called "base-growth". On the contrary, the growth follows a "tip-growth" mechanism when the particle lifts off the substrate (weak interaction) and is observed at the top of the CNTs (Fig. 1.13) [75, 76].

Carbon nanotubes can self-assemble during CVD growth into vertically aligned films or patterns perpendicular to the substrate surface. High density of catalyst particles is needed and then high density of vertically aligned carbon nanotubes is obtained. During CVD growth, nanotubes interact with their neighbours via Van der Waals forces to gain rigidity, which allows the nanotubes to self-orient and grow perpendicular to the substrate [71]. However, in plasma-enhanced chemical vapour deposition (PECVD), the growth direction of nanotubes can be controlled by applying electric fields [70].

PECVD first emerged in microelectronics because certain processes cannot tolerate the high temperatures of the thermal CVD operation. The low temperature operation is possible because the precursor dissociation (necessary for the deposition of all common semiconductor, metal and insulator films) is enabled by the high energy electrons in otherwise cold plasma [77].

The samples are placed onto a heating plate in the centre of the PECVD reactor (Fig. 1.14), which is then pumped down to a low base pressure (~1 mTorr) to evacuate atmospheric gasses. Then the substrate is heated (at 450–700 °C depending on process and chemistry) and the carbon-containing and reacting gases are introduced into the chamber. A high voltage applied to the electrode above the sample causes an ionization of the gases, resulting in plasma formation. The energy from heating the substrate and from the high-voltage plasma causes decomposition of the gas into its components. CNTs grow aligned perpendicular to a substrate surface due to the generated electrical field created by the plasma environment.

Even though a big body of applications has been undertaken with CNTs, there is still a long list of unsolved fundamental issues in the carbon nanotube growth mechanism which make the CNT production with defined characteristics and properties (selective synthesis) rather challenging. For instance, one unfulfilled aspiration is the control over nanotube chirality to obtain pure metallic or semiconducting materials needed for electronics applications (such as interconnects or transistors) or to control diameter and lengths of the CNTs. Such lack of control added to the difficulties in CNT purification, sorting and processing are responsible for the high dispersion of results and performance of CNT devices.

1.1.3 CNT Purification Methods

Purification has been an important effortful task since the discovery of carbon nanotubes; it is a process that separates nanotubes from non-nanotube impurities produced as a result of CNT synthesis. These impurities generally include catalyst metal nanoparticles (iron, cobalt, nickel) together with amorphous carbon and fullerenes that are unavoidably present in the CNT soot.

The lack of a good purification protocol can also constitute an important issue that hampers the applications. Therefore, an exhaustive pretreatment or purification process is required. Generally, the purification methods can be separated into dry methods and wet methods, but usually purification consists of multi-step purification procedures: dry method followed by wet method.

Dry methods of purification refer to methods that can selectively remove, through gas-phase oxidation, amorphous carbon species due to their higher reactivity compared to that of carbon nanotubes. The simplest method is air oxidation at temperatures below the CNT decomposition [78].

Wet methods treat nanotubes in solution for purification purposes. The use of nitric acid (HNO_3) is the most common wet method used as it is straightforward, inexpensive and effective. After repeated filtration steps and maybe also centrifugation steps, purified nanotubes can be obtained [79, 80].

Other procedures rely on the use of either concentrated HNO_3 reflux with H_2SO_4 or HCl, or hydrogen peroxide reflux [81, 82]. Such treatments can eliminate a big amount of residual metal particles and other impurities. As mentioned, most dry methods of purification are also followed by a step of acid treatment to dissolve metal catalyst and/or metal oxides formed during the gas-phase oxidation step [83]. However, and even after prolonged washing treatments, metal catalyst can be still detected on CNT samples [84, 85]. The difficulties in removing the catalyst have been ascribed to the fact that in some cases, the particles are sheathed and protected by graphene sheets. Some groups have claimed that such metal impurities in CNT can cause electrocatalysis on certain electroactive species [50, 86]. In the case of graphene sheathed metal impurities, the redox processes are not taken place in close contact to the metal catalyst, so it is not fully clear how the metal electrocatalysis is taken place when exhaustive purification is performed. One plausible explanation may be related with the possibility that the metal impurity might induce changes in the electronic band structure of the CNT by adding additional energy states which might enhance the electron transfer kinetics.

Table 1.1 Material and electrochemical properties of an ideal electrode

High electrical conductivity
Hard and durable
Homogeneous microstructure throughout the bulk
Reproducible physical, chemical, and electronic properties
Good chemical inertness
Low and stable background current
Morphological and microstructural stability over a wide potential range
Rapid electron-transfer kinetics for a wide range of redox systems
Easily fabricated, shaped, and inexpensive in cost

One of the challenges in achieving high-quality electroanalytical measurements is the reproducible control of the physical and chemical properties of the electrode in order to get low background current and a rapid rate of electron transfer for the target analyte.

1.2 Carbon Nanotubes as Platforms for Electrochemical and Electronic Biosensors

Among all platforms with interest for biosensors, carbon nanotubes can be presented as a groundbreaking material that offers novel possibilities over other systems for the development of novel sensing approaches. In the following section, we will pay special attention to carbon nanotubes and their integration as solid electrodes.

1.2.1 Solid Electrodes

Solid electrodes have been a mainstay in electroanalytical chemistry for nearly five decades now [87]. For a solid material to function as an electrochemical electrode it must possess several characteristics; Table 1.1 lists some desirable electrode properties for electroanalytical purposes.

The most common solid electrode materials used as electrodes for studying electron transfer kinetics and mechanisms for determining thermodynamic parameters are: carbonaceous materials such glassy carbon [88, 89], carbon fibres [90], carbon paste or carbon composite (which consists of graphite particles in contact, incorporated in an inert matrix) [91, 92], carbon black [93], various forms of graphite (powdered graphite, pyrolytic graphite, highly ordered pyrolytic graphite) [94–97] and carbon nanotubes [98, 99]; metals such as gold [100, 101], mercury [102, 103], silver [104] or platinum [105] and semiconductors for example metal oxides [106, 107] and conducting organic salts [108].

The choice of an electrode material depends to a great extent on the useful potential range of the electrode in the particular solvent employed and the qualities and purity of the material [109].

A general advantage of metal electrodes is that their high conductivity results in low (usually negligible) background currents. It is usually fairly easy to increase sensitivity and reproducibility at solid electrodes by forced convection. Their surfaces can be modified by electrodeposition or chemical modification; although the latter is more common with carbon electrodes. Another advantage of the use of metal electrodes is the ease of construction of the electrode assembly, and the ease of polishing.

Of all the solid electrodes utilized in electroanalytical chemistry, carbon materials are probably the most challenging and this is due to the changing surface chemistry depending on the microstructure of the carbon. As a consequence, the electrochemical properties can vary from material to material, but this can also become an advantage.

Carbon [110] exists in various conducting forms. An important issue related to surface reactivity is the available potential range of carbon electrode materials over which background reactions contribute negligibly to the observed current. Such reactions depend strongly on the nature of the carbon electrode material as well as the preparation of the surface [38]. The kinetics of surface oxidation and hydrogen evolution are significantly slower on carbon than on most commonly used metal electrodes, and the resulting wide potential window is one reason for the widespread use of carbon materials for electrodes [37].

Moreover, carbon materials have significantly more complex chemistry than metals. Carbon has a high surface activity, which explains its susceptibility to being poisoned by organic compounds. Bonds with hydrogen, hydroxyl and carboxyl groups, and sometimes quinones, can be formed at the carbon surface. The presence of these groups signifies that the behaviour of these electrodes can be very pH-sensitive, but this has also been purposely used to modify the electrode surface. That is, these groups provide sites for covalent linking to biorecognition elements (or other materials) or for their integration onto polymer surface structures, key points for the development of biosensors.

Thus, the advantages of carbon electrodes include low cost, wide potential window, relatively inert electrochemistry, richness in surface chemistry and electrocatalytic activity for a variety of redox reactions.

Among all carbon conducting forms, CNTs exhibit the mentioned remarkable electrical, chemical, mechanical and structural properties that make them a versatile material for the design of unusual electrode configurations.

1.2.2 CNT Platforms

In order to guarantee an efficient performance of the CNT based electrochemical devices, attention has to be paid not only to CNT synthesis and purification but also to the way that the CNT electrode is built up. In this section, the different carbon nanotube deposition techniques and carbon nanotube arrangements on different electrode surfaces will be described: on one side, randomly dispersed CNT in a polymer matrix or the spaghetti-like CNT dispersed on conductive surfaces; and on the other side, CNT arrays grown in situ on substrates with large scale control

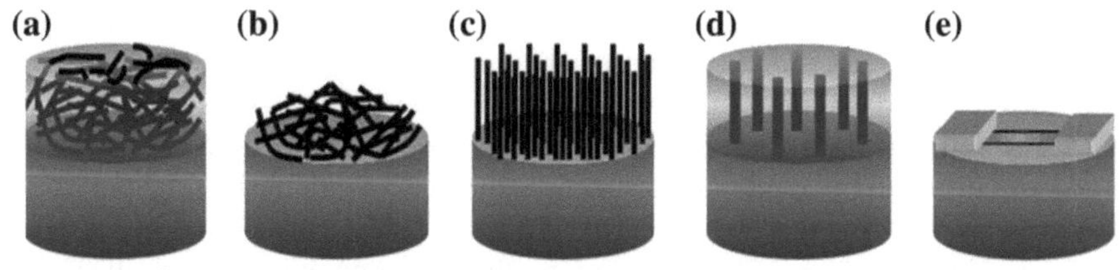

Fig. 1.15 Illustration of different CNT electrode configurations: from *left* to *right*, **a** randomly dispersed CNT composites, **b** randomly dispersed CNTs on a surface, **c** vertically aligned CNTs (VACNTs), **d** oriented CNTs embedded in a polymer matrix, **e** individual CNT electrode (CNT-field-effect transistor)

of location and orientation, embedded or not in a polymer matrix. Finally, individual CNT electrodes will be briefly discussed. Figure 1.15 illustrates these different CNT electrode arrangements.

1.2.2.1 Rigid Carbon Nanotube Composite

Composites result from the combination of two or more dissimilar materials. Each individual component keeps its original nature while giving the composite distinctive chemical, mechanical and physical qualities, different from those shown by the individual components [111]. From this definition, due to the conducting or insulating character and to the nature and arrangement of each component, composites comprise a wide scope ranging from metal/ceramic composites, polymer-based composites to natural or biomimetic biocomposites. Among the broad world of composites, the conducting ones attract special attention because of their special applications in electronics and electrochemistry. Composites for electrode applications are formed by at least one conducting and one insulating phase. In this context, carbon-based materials arise as ideal conducting phase in composites used for electrochemical sensors. One of the more practical uses of carbon based electrodes is in a form of a resin composite. In general, carbon composite materials present improvements in the electrochemical response of the conventional solid carbon electrodes. They can exhibit the interesting advantages of high mechanical stability, easy surface renewal without extensive polishing, lower background current and in some cases, microelectrode array behaviour. The later aspect, brings about even more remarkable high signal/noise ratio, fast electrode response, low detection limits and low cost. The last point makes them ideal for mass produced sensors, especially when one compares them with other materials such as gold and platinum [112].

The electrical properties of the composite will therefore depend on the nature of each of the components, their distribution, and also on their relative quantities. The electrical resistance is determined by the connectivity of the conducting particles inside the polymeric matrix. This means that the relative quantity of each composite component has to be studied in terms of percolation curves [113] to determine the optimal composition. Different composite material based on graphite powder embedded in diverse kinds of polymeric matrices have been developed in the last few decades, obtaining interesting electrochemical applications such as in the enzymatic, DNA-detection, immunosensing field, etc. [114–117]. Nowadays, high interest is focused on composites based on carbon nanotubes (CNTs).

Many soft and rigid composites of carbon nanotubes have been reported [42]. They are formed by randomly dispersion of CNTs with insulating polymeric matrices such as granular Teflon [118], chistosan [119], polystyrene [120], polysulfone [121] and epoxy [122, 123] or by their incorporation into a silicate gel matrix [124, 125]; or with conductive polymers such as polypyrrole [126] and polyaniline [127] which can be fabricated by chemical or electrochemical polymerization [42, 124, 128, 129]. Irrespective of the nature and the way in which the composite is built up, these matrices are important for providing robustness to the CNT platform and at the same time for facilitating the analyte trapping which can result quite advantageous for some applications in the (bio)chemical sensing field. They also exhibit the valuable advantage of generating a fresh surface by only polishing every time that is needed for successive experiments.

Note that the use of insulating polymers can be more advantageous than the conductive polymers when employed in electrochemistry (e.g. cyclic voltammetry). This is because the former electrodes exhibit a more discernible response for the redox species from the background current [124].

Motivated by these attractive properties, part of this work focuses on the transduction capabilities of rigid carbon nanotube-epoxy composites to give more highlights to a previous and brief study of such CNT-epoxy systems as platforms for electrochemical sensing [123]. These rigid carbon nanotube composites constitute a simple and cheap electrode material where the insulating polymer (epoxy resin) can help to reduce the background signal thus enhancing the signal noise ratio.

Some drawbacks in CNT composites reside in the CNT itself, such as in the lack of homogeneity of the different commercial CNTs lots due to different amount of impurities in the nanotubes, as well as dispersion in their diameter/length and state of aggregation (isolated, ropes, bundles). These variations are difficult to quantify and can affect the final properties of the CNT-composites.

1.2.2.2 CNT/Glassy Carbon

Both SWNTs and MWNTs can be deposited directly from CNT dispersions as a random network or thin film on conventional electrodes. From the point of view of their construction, such electrodes are very easy to prepare but they may suffer from mechanical instability, thus limiting their application. A typical electrode substrate for building up non-oriented CNT films is glassy carbon.

Glassy carbon (GC), also referred to as vitreous carbon, is widely used as a solid carbon electrode for electroanalysis [109]. Unlike HOPG and other graphites, GC is hard and microstructurally isotropic; the interwoven sp^2 carbon ribbon gives rise to its mechanical hardness, so the material is polishable and impermeable to gases and liquids. The density of the GC is less than the HOPG one, indicating that the material contains some void space (nanoporosity) [130]. Moreover, glassy carbon is often used as a support material because of its electrochemical inertness in a wide potential window, its amorphous structure with a random distribution of active sites for nucleation [131], its mentioned easy regeneration, robustness and low cost [132, 133].

The modification of glassy carbon is currently done by casting drops of purified CNTs dispersed either in organic or aqueous solvents and allowing the electrode to dry [134–136]. Glassy carbon provides a hydrophobic surface which considerably stabilizes the CNT films.

Another carbon substrate widely employed is the basal plane pyrolytic graphite [46, 47, 49]. In this case, carbon nanotubes are abrasively immobilized onto basal pyrolytic graphite electrodes by gentle rubbing the electrode surface with a fine quality filter paper containing CNTs.

However, in some cases and even more when the CNT layer is too thin, the underlying glassy carbon or basal pyrolytic graphite electrodes can provide some interferences in the electrochemical response. On the other side, very thick 3D CNT films can contribute with large capacitive background currents due to the large surface area which can hamper electroanalytical applications, as often occur when using the highly porous bucky paper [122, 137]. In such CNT papers, formed basically by filtration of purified SWNTs, a small faradaic signal can be lost in the high capacitive currents. However, such high CNT surface area systems result beneficial for energy storage or electrochemical actuator applications.

Gold or platinum electrodes have also been used as conducting supports for developing CNT modified electrodes. Although an enhanced electrochemical response can be evidenced, the CNT films (when casting from CNT dispersion) are not very stable on such hydrophilic surfaces and tend to be removed in contact with aqueous media, a problem that can be solved by covalent CNT attachment to the gold surface through thiol linkages.

1.2.2.3 Vertically Aligned Carbon Nanotube Electrodes (VACNT-Microelectrode/Macroelectrode)

For a thorough understanding of the fundamental CNT electrochemistry, many applications require well-organized and oriented carbon nanotube arrays. For instance, up-right CNT configurations (e.g. perpendicular to the substrate) can provide a high density of oriented electroreactive edges facilitating, on one side, the electron transfer process, and on the other side, the covalent linkage of molecules.

There are two major approaches for obtaining oriented CNTs: by self-assembly procedures (many groups have reported the assembly of shortened SWNTs in an orthogonal orientation on a gold electrode modified with thiol self-assembled monolayers with –NH_2 terminal groups which can attach the CNTs via carbodiimide chemistry) and by in situ growth of aligned CNTs with plasma-enhanced chemical vapour deposition (PECVD). However, by applying thermal CVD processes, high density of vertically aligned carbon nanotubes can be achieved if a dense enough distribution of the catalyst particles is deposited.

Some groups have exploited these methodologies and have developed, for instance, vertical 3D CNT ensembles with different CNT density embedded or not in an insulating matrix (SiO_2) [137–139]. These CNTs are grown directly onto an electrode surface (e.g. Si coated with a Cr film) by using plasma enhanced-CVD [137–140].

In our approach, we will use commercial macroelectrodes (or forest CNT macroelectrodes) of high-density multi-walled carbon nanotubes (about 0.2 g cm^{-3}) grown on a silica substrate by chemical vapour deposition and microelectrodes composed of patches of 300 μm $\times$ 300 μm with high density of vertically aligned carbon nanotubes (also grown by thermal CVD process) which have been developed in the National Microelectronics Centre (CNM, Barcelona) [141].

1.2.2.4 CNT-Ultramicroelectrode Arrays

Arrays of CNTs can provide more benefits from the electroanalytical point of view if their geometrical disposition is perfectly controlled in order to make them act as nanoelectrode arrays [138, 139]. That can be accomplished with the "opal inversion" technique which will be discussed in this section, for the development of CNT-ultramicroelectrode arrays (CNT-UMAs).

The discovery of the unusual properties of the ultramicroelectrodes in the eighties opened new electrochemical and analytic possibilities, and new schemes of detection that were not possible with conventional electrodes [142–145]. Ultramicroelectrode is a term used to describe microelectrodes where at least one of their dimensions is smaller than the thickness of the target analyte diffusion layer [146]. Basically, these types of electrodes exhibit a fast non lineal diffusion as the dominant way in the mass transport. When the electrode is miniaturized, the diffusion of the species on the surface of the electrode is predominantly radial, for which the species are driven also axially increasing significantly the mass transfer rate. Quickly a steady state current, independent of the convection, is reached which is directly proportional to the analyte concentration. That confers to the electrode a very fast response, very high signal/noise (faradaic current/non faradaic charging current) relation, and consequently very low limits of detection, all these properties highly desired in amperometric sensors.

Other benefits that present these ultramicroelectrodes are their low ohmic potential drop and their application in highly resistive media. However, the current of an individual ultramicroelectrode is very low and extremely sensitive instrumentation is required for its detection [142–145]. An alternative way to maintain the electroanalytical advantages of the individual ultramicroelectrodes but with a response of amplified current is by means of arrays of ultramicroelectrodes (UMAs). These arrays may consist of multiple identical microelectrodes connected in parallel with a well-controlled and uniform interdistance among each ultramicroelectrode or in a more random distribution. From the theoretical point of view, the signal from these devices is amplified as many times as the number or microelectrodes that defines the UMA [147]. The state and magnitude of the diffusion will depend mainly of the diameter of these microelectrodes and of the

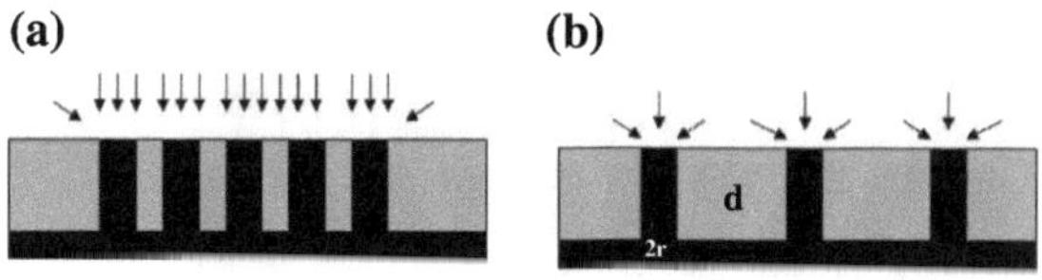

Fig. 1.16 Packaging density of microelectrodes: **a** *M* Macroelectrode behaviour, **b** *UMA* Ultramicroelectrode array

distance among them, by which the optimization of the most adequate geometry according to the requirements is indispensable.

It has been assessed that loosely packed arrays where the inter-electrode distance d $\gg$ 2r (r being the radius of a single microelectrode) yield the expected current signal (*m* times amplified) whereas closely packed arrays, where d $\approx$ 2r, behave as a macroelectrode having a current that is proportional to the total geometric area of the microelectrodes in the array [148] (Fig. 1.16).

Therefore, one of the big challenges in this area is the preparation and production of these ultramicroelectrodes. The most utilized methodologies are based on the use of inert templates in which metallic microelectrodes can be dispersed [149, 150], serigraphic processes of thick layer ("screen printing") [151, 152] or lithographic techniques of thin layer employed in the fabrication of the microelectronic circuits. With these techniques individual miniaturized configurations in the shape of disks, bands, fine wires, rings, or forming a matrix or interdigitated structures could be obtained [142–145]. While the two first techniques are more versatile with extensive range of applications, cheap and open to a larger scientific community, the last one needs more expensive equipment but presents the advantage of a greater control in the miniaturization of the systems, high reproducibility and greater massive production. The microelectronic technique also allows integrating in a same chip the transducers with their complementary electrodes (reference and auxiliary electrodes) and the corresponding circuit lines for the signal processing. Apart from that, the microelectronic technology permits the production of microelectrodes of different nature in a same chip (Fig. 1.17), which can be individually addressed in an electrochemical experiment. This type of system results very appropriated for complex samples with more than one analyte since it allows working with different ranges of sensitivity and selectivity and, therefore, to enlarge the precision of the measurement. Additionally, if the assembly is constituted by microelectrodes of different material, a greater specificity of the multisensor may be devised.

An interesting alternative to the lithographic techniques of the microelectronics is the one proposed in this thesis work. This technique allows the development on CNT-UMAs based on the concept of self-assembly of an inert template and on the deposition of the catalyst material for CNT growth by different physicochemical processes. Once the template is removed, a nanostructured pattern of catalyst seeds is achieved. This allows, later on, controlling the geometric disposition of the transducer (CNTs). The technique, being cost effective, constitutes a relatively simple one and of great accessibility that takes advantage of the capacity of self-organization of some materials that form highly ordered structures on surfaces. As an example of these types of materials, we can mention the microspheres of polystyrene, which dispersed on a substrate, tend to form structures with hexagonal packaging [40]. These materials

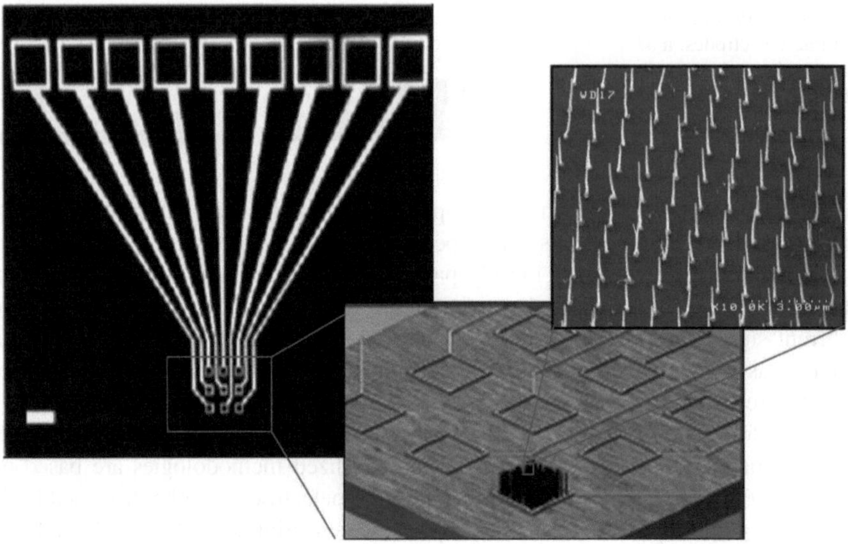

Fig. 1.17 Use of the microelectronic technology. From *left* to *right*, 3 × 3 individual nanoelectrode array with accompanying contact pads; CNTs grown using PECVD; SEM image of a zoom of the array showing CNTs growth. Adapted with permission from Ref. [153]

turn out to be very versatile as templates for the formation of porous patterns or highly periodic micro/nanostructures by means of the "opal inversion" technique.

Electrochemical Properties of High Density Versus Low Density of CNTs

To understand the electrochemical behaviour of such systems, it is worth showing an example where the response of ultramicroelectrodes developed by means of lithographic techniques is compared with the macroelectrode one [138]. The system UMAs requires the spacing among individual nanoelectrodes to be sufficiently larger than the diameter of the nanotubes to prevent the overlap of diffusion layers from the neighbouring electrodes. Figure 1.18 shows different density CNT arrays and their corresponding electrochemical response. At high density of CNTs, the electrochemical response is dominated by a planar diffusion transport evidenced by the common peak shape in the cyclic voltammetry; whereas at low density of CNTs, the electrochemical response is basically sigmoidal which is typical of radial diffusion mass transport. Thus, low density CNT nanoelectrode arrays displays the desired independent nanoelectrode behaviour showing diffusion-limited steady-state currents in cyclic voltammetry over a wide range of scan rates [138].

Therefore, the size and the spatial distribution of CNTs can be precisely controlled to ensure that each one behaves as an independent nanoelectrode. We accomplish this, as mentioned before, with "opal inversion" technique. So, this configuration can be very promising for developing ultrasensitive electrochemical sensors.

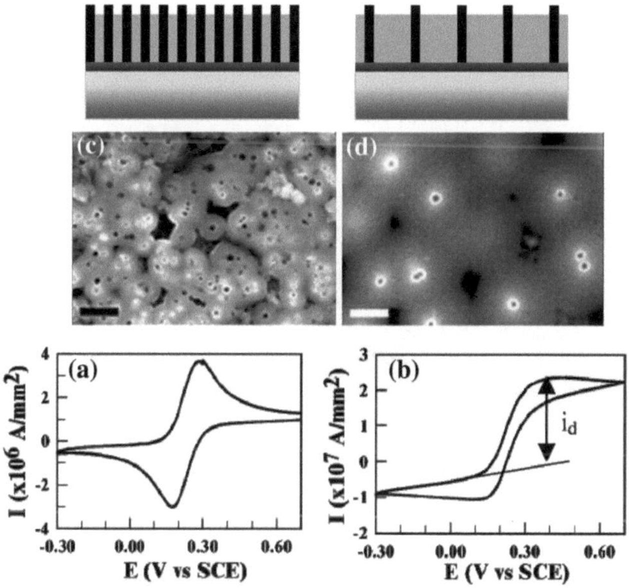

Fig. 1.18 **a, b** CV measurements in 1 mM of Fe(CN)6 $^{3-/4-}$ and 1 M KCl with the high-density MWNT nanoelectrode array (2×10^9 electrodes cm^{-2}) and low density one (with ca. 7×10^7 electrodes cm^{-2}) respectively; **c, d** show the scanning electron images for the high and low density arrays. Readapted with permission from Ref. [138]

1.2.2.5 Carbon Nanotube Field-Effect Transistor

An electrochemical sensor response can be accomplished by using not only the common electrochemical configurations but also devices based on electronic detection in a field-effect transistor configuration (FET) [154]. We have already addressed this device when discussing the electronic properties of the CNTs. At that time, we pointed out that this kind of system allowed probing the electronic properties of the CNT and discriminating semiconducting nanotubes from metallic ones.

The first demonstrations that semiconducting single-walled carbon nanotubes could be used as conducting channels in FETs with switchable properties appeared in 1998 [155, 156]. As stated before and shown in Fig. 1.19, such configuration consists of a semiconducting single-walled carbon nanotube or a network of semiconducting SWNTs which are contacted by two electrodes, the source (S) and drain (D) electrodes. There is also an additional electrode which is separated from the CNT by an insulator film, gate electrode (G). A voltage applied between source and drain allows a current flow in the conducting carbon nanotube which can be modulated by applying a voltage on the gate electrode. Thus, the role of the gate voltage is to control the charge carriers in the current flow. The gate voltage can induce accumulation of holes (electrons) in the channel if it is polarized

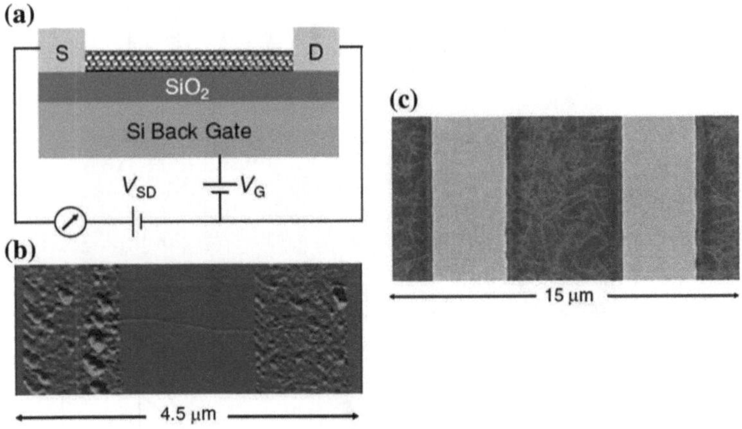

Fig. 1.19 **a** Typical scheme of a CNT-FET with the different electrodes in the so-called backgate configuration: source (*S*), drain (*D*) and the backgate electrode (*G*) (doped silicon), SiO_2 is the dielectric or insulator layer; **b** atomic force microscopy image of a typical CNT-FET device with individual SWNTs connecting the (*S*) and (*D*) electrodes; **c** scanning electron microscopy image of a FET device consisting of a random array of CNTs. Adapted with permission from Ref. [154]

negatively (positively) which makes the valence and conduction bands to move upwards (downwards) relative to the Fermi level.[6]

A simplistic picture for describing the flow current through the nanotube is by matching the valence or conduction band of the nanotube with the Fermi energy (E_f) (Fermi energy is the energy of the highest occupied state at the absolute zero temperature.) of the contact electrodes through the effect of the electric field provided by the gate voltage. However, the electric field generated by the gate voltage which controls the current flow through the CNT is dependent not only on the electronic characteristics of the nanotube itself (that is on their band-gap characteristics) but also on the interface between the SWNTs and the metal electrodes where Schottky barriers (SB)[7] can be formed (Fig. 1.20). Schottky barriers can be generally formed when a semiconductor is brought into direct contact with a metal due to the mismatch of the work functions[8] of the metal and the CNT. A charge

---

[6] It can also be seen as if the Fermi level were shifting upwards or downwards relative to the conduction and valence bands with electron or hole doping respectively.

[7] Schottky barrier is a potential barrier formed if a semiconductor is brought into direct contact with a metal. A charge transfer occurs until the Fermi levels of the metal and the semiconductor are aligned. The charge transfer creates a depletion zone in the semiconductor near the junction (electrical dipole). Such depletion causes a rectifying behaviour. Rectifying means that carrier transport is hindered in one of the two possible current directions. Not all metal–semiconductor junctions form Schottky barriers. A metal–semiconductor junction that does not rectify current is called an ohmic contact. Rectifying properties depend on the metal's work function, the band gap of the intrinsic semiconductor, the type and concentration of dopants in the semiconductor, and other factors.

[8] The work function is the minimum energy needed to remove an electron from a solid to a point immediately outside the solid surface (or energy needed to move an electron from the Fermi level into vacuum).

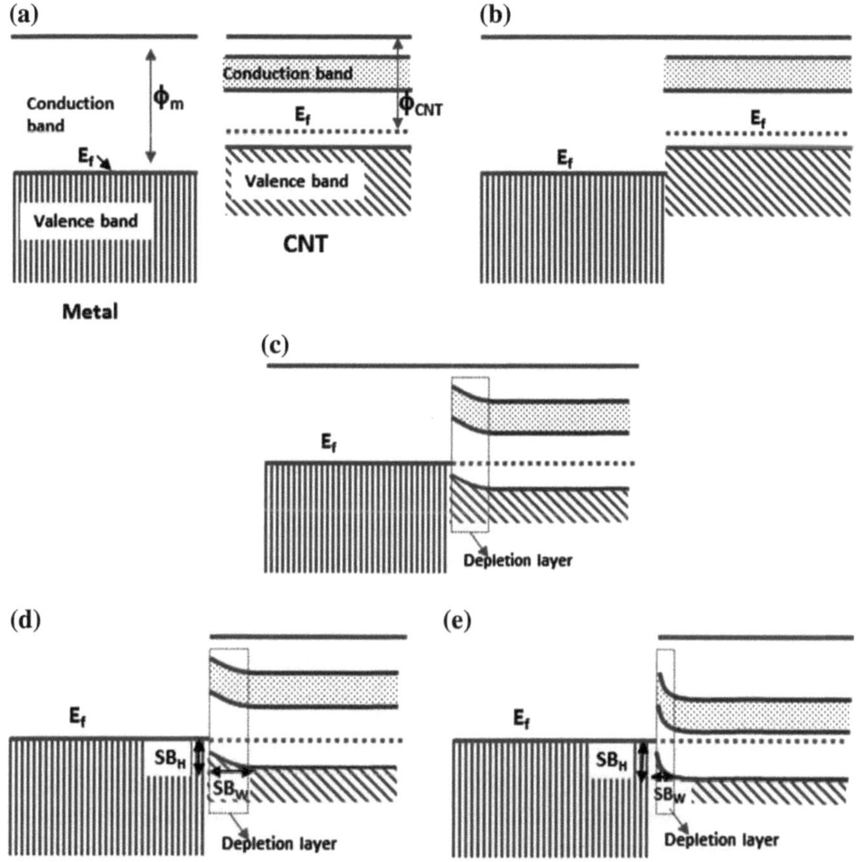

Fig. 1.20 Scheme illustrating the formation of Schottky barriers (*SB*). In (**a**) the band diagrams for the metal contact and for a semiconducting p-type nanotube are shown. The metal has a higher work function or barrier height (ϕ) than the CNT. The CNT has been schematized as a p-type since oxygen from the environment can dope the CNT with holes. In this case, the Fermi level of the CNT is close to the valence band. When the CNT and the metal contact are brought together (**b**), the Fermi level of both materials will try to equilibrate by flowing charge carriers between them (**c**). That results in a bending of the bands forming a depletion layer which gives place to the generation of Schottky barriers that inhibit the charge carrier transport. However, the height (SB_H) and thickness (SB_W) of the barrier can be changed with the gate voltage. In (**d, e**) the height and the thickness of the barrier height are indicated. It also shows how the thickness of the tunnelling barrier can be decreased with the gate voltage. Generally, in CNT-FETs the charge transport at the Schottky barriers is carried out by tunnelling processes and therefore for allowing the current flow the barrier thickness has to be decreased with the gate electrode voltage

transfer occurs at the interface which results in the bending of the conduction and valence bands. Such bending generates a charge depletion (potential barrier) that opposes further charge transfer. There are two of such barriers, one at the source electrode and the other at the drain one and can dominate the transport characteristics of the CNT. The thickness (SB_W) and height (SB_H) (Fig. 1.20d) of such

Fig. 1.21 p-type CNT-FET device characteristics: **a** threshold voltage, **b** conductance modulation, **c** maximum conductance, **d** hysteresis, **e** transconductance

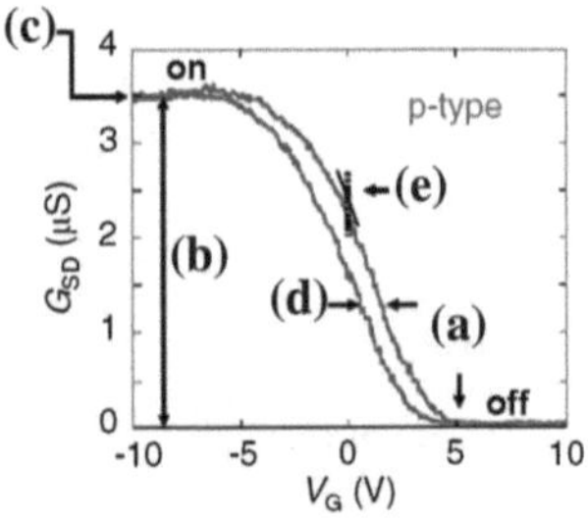

barrier can be increased or decreased with the electric field provided by the gate electrode and thus modulating the current flow either of holes or electrons. In general in CNT-FETs the flow current is more dependent on the thickness than on the height of the barrier. Mechanisms based on tunnelling processes and hence on the thickness of the barrier are more important at the contact/CNT interface than the ones depending on temperature which are more dependent on the height of the barrier. Therefore, when the Schottky barrier becomes thinner, tunnelling of the charge carriers can be produced favouring the current flow in the CNT-FET.

So far we have stated that the charge transport in a CNT-FET is not only determined by the electronic characteristics of the semiconducting CNT itself but also by the interfacial Schottky barriers which can be modulated with the gate electrode. Next, we will illustrate the typical features that characterize the CNT-FET response.

In Fig. 1.21 the basic features of the so called "transfer characteristic" of a FET device are shown for a p-type CNT-FET. Applying a variable gate voltage (V_G) through the gate electrode, the source-drain current (I) or conductance (G) is measured at a fixed source-drain voltage (V_{SD}). In the plot (a) represents the threshold voltage which corresponds to the minimum voltage that is required to switch on the conductance of the tube, (b) represents the modulation (considered as the variation of the conductance from the threshold voltage), (c) represents the maximum conductance, (d) illustrates the device hysteresis and (e) indicates the transconductance[9] which is the relation between the current change as a function of the gate voltage change and provides information of the charge carrier mobility.

In this case, the CNT-FET is totally switched off for the positive voltages, a typical behaviour for a p-type system in which the holes are the dominant charge carriers. As said, most CNT-FETs exposed to air show p-type behaviour but become more ambipolar in vacuum. Such doping effect can take place during the synthesis and handling of nanotubes or by oxygen adsorption on the CNT-metal interface when exposed to air or by charge transfer from high work function contact electrodes. For instance, the oxygen changes the work function of the metal contact compared to its value in vacuum shifting the energy of the Fermi level of the metal with respect to the valence and conduction band edges in the CNT and thereby

[9] The transconductance, g_m, can be expressed as $g_m = C_t|V_G - V_{thr}|\mu/L$, where C_t is the total capacitance, V_G is the gate voltage, V_{thr} is the threshold gate voltage, μ is the charge carrier mobility and L is the length of the tube.

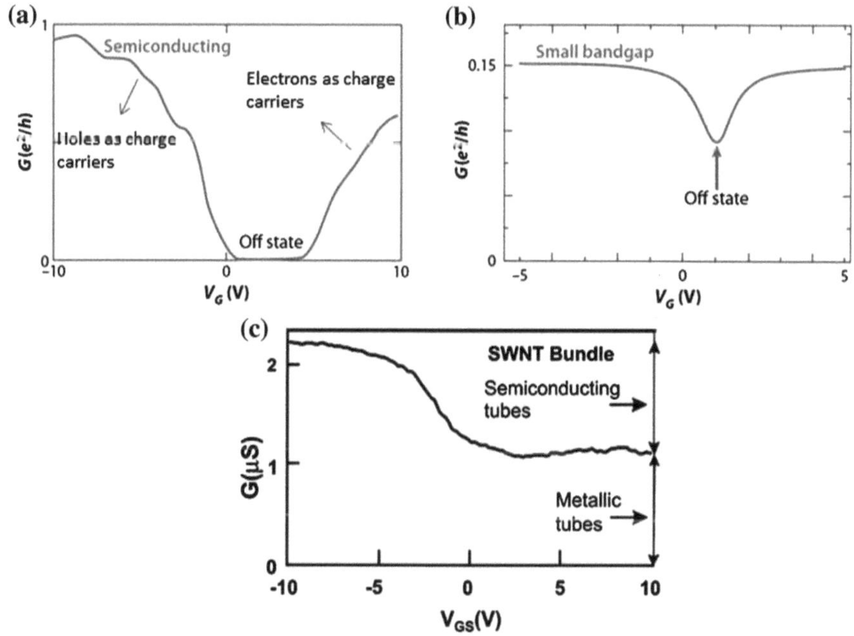

Fig. 1.22 CNT-FET response with **a** ambipolar characteristics, **b** when the CNT has also an ambipolar behaviour but with a small band-gap, **c** when CNT network contains metallic tubes

altering the hole and electron Schottky barrier heights. This mechanism has been demonstrated when removing oxygen from a CNT-FET by annealing in vacuum. The transfer characteristics of the device gradually changed from p-type to ambipolar as the oxygen was removed [157]. Thus, CNT-FETs can exhibit different features depending on their own characteristics and their environment and Fig. 1.22 collects some of them. The first example represents a CNT-FET with ambipolar characteristics. Ambipolar behaviour is observed when the Schottky barriers for holes and electrons are similar and the current flow can be done by holes or electrons. Therefore, high currents for both negative and positive gate voltages and different on/off current ratios are expected depending on the magnitude of the CNT band-gap. In principle, the CNTs are naturally ambipolar, they can transport holes and electrons. However, and as mentioned before, depending on the CNT processing, the metal contacts and their effects on the Schottky barriers, the CNT-FET can sometimes act as a unipolar device and preferentially transport only one type of carrier. The figure also shows the device characteristics when a CNT-FET has a small band-gap. In this case, a very poor on/off current ratio is also expected. Finally, the figure additionally shows the response of a CNT-FET device formed from a CNT network in which metallic tubes are present. In such case, the G–V_G signals cannot be totally switched off due to the fact that metallic tubes cannot be modulated with V_G.

Following, we will illustrate how the band diagrams of the metal contact/ CNT are bent under full device operation, that is with a fixed S/D voltage and

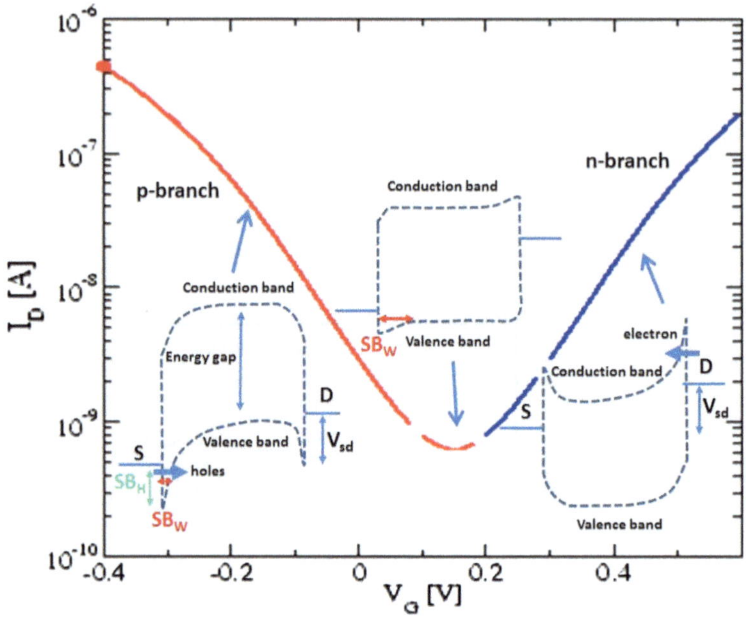

Fig. 1.23 A typical conductance versus gate voltage profile for an ambipolar CNT-FET. The *plot* shows the p-branch, at negative voltages where the hole charge carriers are dominant in the current flow and also the n-branch at positive gate voltages where the electrons are the dominant charge carriers. The figure also depicts how the band diagrams (valence and conduction bands are bent at the different gate electrodes producing Schottky barriers at the source (*S*)/CNT and drain (*D*)/CNT interfaces. The curve has been carried out by applying a fixed voltage between *S* and *D* (S_{SD}). The schematic band diagrams also show the main charge transport at the different branches: high tunnelling of holes from the source in the p-branch at which the Schottly barrier is thin; low charge transport due to thick Schottky barriers at intermediate gate voltages; high tunnelling of electrons from the drain at which the Schottky barriers become thin again. Note that current flow of electrons between the contacts takes place from higher energy to lower energy (e.g. the Fermi level of the drain is at higher energy than the one of the source), the current flow of holes takes place on the other sense from lower energy to the higher one. The scheme also indicates the Schottky barrier height (SB_H) and thickness (SB_W)

a variable gate voltage. In the Fig. 1.23, the conductance of an ambipolar CNT-FET device is shown together with their respective band diagrams for different gate voltages. As the gate electrode is negatively biased, positive charge carriers (holes) are attracted towards the CNT and accumulate in it. This additional charge causes the valence and conduction bands to move up relative to the Fermi level in the CNT. That makes the Schottky barrier for the holes at the source become thin enough to allow for a considerable amount for tunnelling. For intermediate gate voltages, the barrier is too thick for tunnelling and the conductance in the CNT device decreases. At high positive gate voltages, the bands bend down making the Schottky barrier for electrons at the drain contact thinner, giving rise to an electron tunnelling current.

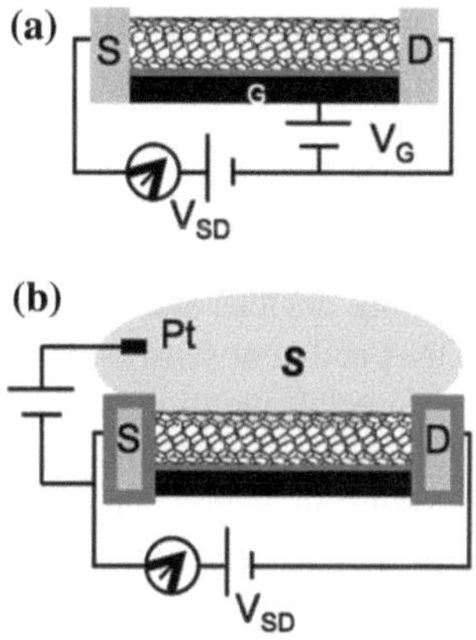

Fig. 1.24 Comparison schemes of **a** CNT-FET in the backgate configuration: source (*S*), drain (*D*) and the backgate electrode (*G*); **b** CNT-FET in liquid configuration: passivated source (*S*) and drain (*D*) electrodes, Pt wire (pseudo reference electrode), solution containing analyte (*S*)

A very important point to highlight with respect to the CNT-FET devices is that the CNT conducting channel exhibits some advantages with respect to conventional silicon based FETs and which makes attractive for sensing purposes. The nanotube is located on the surface open to the environment (contrary to the conventional silicon FETs in which the conducting channel is buried in the bulk material). Moreover, and due to the tubular structure of the tube, the current flow takes place at its surface in direct contact with the environment. They also conduct ballistically, which could lead to faster devices. All these features make CNT-FETs be very sensitive to tiny changes in the surrounding environment and can become promising candidates for the development of ultrasensitive (bio)chemical devices.

We have mentioned that in general, the gate electrode is a doped silicon substrate at the back of the device (backgate electrode). However, in other circumstances it is appropriate to perform electronic detection in aqueous environment, which is the more suitable environment for monitoring biomolecules or biological processes. Figure 1.24 schematizes the difference between both systems. Under aqueous environment, the electrolyte can induce gating effects in an alternative way to the conventional back gating configuration by using an electrode immersed in a solution [154]. In this configuration, a reference or pseudoreference electrode (usually Ag/AgCl wire or Pt wire respectively) is placed in contact with a solution containing the analyte which at the same time is in contact with the SWNTs. The electrostatic potential difference between the solution and the CNTs is controlled through the gate voltage and the potential at the metal–liquid interface. In such liquid environment, the bio-analyte reaches the CNT surface altering locally the electrostatic conditions or inducing charge transfer at the CNT-liquid interface. In this configuration, passivation of metal contacts is required for minimizing electrochemical side reactions; thus, limiting the sensing region only to the carbon nanotube [154]. It has been found that the CNT-FET configuration in liquids increases the device sensitivity; nevertheless, the main disadvantage is related to the stability and miniaturization of the liquid gating electrode and/or conditions of the electrolyte [158].

1.3 Electrochemical and Electronic Applications of CNT Electrodes

In this section, we will describe the applications of CNTs when used as electrode platforms for instance as electrochemical actuators, as electrochemical energy-harvesting devices and as biosensors (enzymatic and redox protein biosensors, CNT/DNA and genosensors, immunosensors, CNT aptasensors).

Special attention will be paid on biosensors and related issues: the immobilization methods of the biorecognition element onto the electrode surface and the biosensing events that take place on the different CNT platforms together with the mechanism of action of some biological recognition elements.

1.3.1 Electrochemical Actuators and Electrochemical Energy-Harvesting Devices

Many applications rely on the direct conversion of electrical energy to mechanical energy through a material response. The more common materials for electromechanical actuators are represented by electrostrictive[10] materials. However, applications are restricted by the maximum allowable operation temperature, the need for high voltages, and limitations on the work density per cycle [159]. CNTs appear as a new class of electromechanical actuators which generate higher stresses than natural muscle. Like natural muscles, the macroscopic actuators are assemblies of billions of individual nanoscale actuators. The first CNT actuator developed was a SWNT buckypaper which produced elongation/contraction because of the change in dimensions of the nanotube in the covalently bonded direction caused by charge injection with an applied electric potential (electrochemical doping) [159]. Other CNT actuators have been reported [160, 161], but there are still many challenges to overcome in order to develop tailored practical devices mainly because of the uncontrollable properties of nanotube buckypaper and polymer CNT nanocomposite actuators.

Moreover, carbon nanotubes can be employed either as electrode materials or conductive fillers for the active materials in various electrochemical energy-storage systems [5]. The design of 3D CNT ensembles with high surface area and mesoporous character (graphitized structure with internal channels) can yield large double layer capacitances, or provide suitable electrode frameworks for ion intercalation or catalyst entrapment/deposition, all being important for the development of electromechanical systems or electrochemical energy storage systems. Therefore, for energy generation and storage, nanotubes hold promise as supercapacitors, Li-ion batteries (LIBs), fuel cells and solar cells due to their high specific surface area and mesoporous character and their electronic and optical properties [5].

[10] Electrostriction is a property of all electrical non-conductors, or dielectrics, that causes them to change their shape under the application of an electric field.

1.3.2 Biosensors

One of the cornerstones of CNT applications in electrochemistry lies in the development of (bio)sensors. It has been pointed out the outstanding electrochemical properties of CNTs as a consequence of their electronic band structure and their richness in surface structure and chemistry. As mentioned before, CNTs have revealed in many cases electrocatalytical properties accompanied by enhanced electron transfer rates, increased signal currents and decreased overpotentials. Although these properties have been demonstrated to be not so different from the ones corresponding to edge HOPG, the main advantage in the use of electrochemical CNT devices comes from their nanometre size, their high surface area and their versatility to tailor the CNT surfaces in different geometrical arrangements in order to expose the more electroactive sites and facilitate the electron exchange. Another advantage comes from their chemical anisotropy and their interesting covalent/non-covalent chemical functionalization possibilities, many of them facilitated by their curved surface, which can be used for selective chemical modification in the development of electrochemical and electronic biosensors.

1.3.2.1 Immobilization Methods onto the Electrode Surface

In this part we will talk about the different immobilization methods of the biorecognition element onto the electrode surface. As we will see, the chemical and biochemical modification of CNTs possesses special interest because it opens the door to the achievement of novel hybrid systems for functional devices by combining the unique properties of CNT with the ones resulting from functionalization, for instance with complexing agents, fluorescent and electroactive groups, (bio)catalysts, or biomolecules (proteins, carbohydrates, nucleic acids, etc.).

Chemical and Biochemical Functionalization

Chemical modification turns out to be one of the key issues for the development of selective electrochemical (bio)sensors for (bio)recognition or catalyst materials to be used in energy-storage devices. Moreover, chemical modification also aids to overcome one of the drawbacks for CNT applications which is the difficulty of dispersing them in solvents, especially in aqueous media.

As produced CNTs are tightly bundled in ropes (through Van der Waals interactions) which make them insoluble in aqueous and organic solvents, thus compromising their processability. Additionally, and taking into account the lack of control over CNT size and chirality during synthesis, chemical modification is becoming a very useful tool for post-produced CNT sorting.

The chemical modification of CNT can be endohedral (inside the cavity of the tube) or exohedral [162]. There are some examples in literature which have demonstrated the filling of CNT with fullerenes, biomolecules (proteins, DNA), metals and oxides which have been driven inside by capillary pressure [36, 162–169]. However, this thesis work has been focused on exohedral functionalization, taking place just at the external walls of the tubes. Both covalent (chemical bond formation) and non-covalent (physiadsorption) functionalizations can be carried out. For electrochemical devices, such functionalization schemes can be performed either before or after the CNTs are assembled on the electrodes.

Covalent Modification

As mentioned before, the CNT ends (either capped or as open edge ends) are chemically more reactive than the CNT walls. When the ends are open, the unsatisfied valence bonds become very reactive in presence of water or oxygen and normally they are found enriched in oxygenated functionalities. Additionally, the capped ends can be opened and enriched with oxygenated species by oxidizing treatments with chemical agents (e.g. HNO_3), oxygen plasma or electrochemical anodization. These treatments can also introduce oxidized points in the sidewalls [162]. Among the oxygenated functionalities, one of special interest is the carboxylic group. These –COOH residues constitute useful sites for further chemical functionalization through amide or ester bonds. The scheme in Fig. 1.25 shows the covalent bonding of primary amine molecules through amide bonds facilitated by the carbodiimide chemistry which is aided by activated reagents such as N-hydroxysuccinimide (NHS) and dicyclohexylcarbodiimide (DCC) for organic solvents or EDC (1-ethyl-3(3-dimethyl amino propyl)carbodiimide hydrochloride) and sulfo NHS (N-hydroxysulfo-succinimide) for aqueous coupling [10, 34, 170, 171]. Amines can also be coupled to carboxyl groups activated with $SOCl_2$. Ester linkages taken place by reacting the carboxyl-residues with alcohol functionalities are also shown in Fig. 1.25 [10, 170]. The amide formation is being widely used for the anchoring of proteins, amine terminated oligonucleotides, ion receptors, gold nanoparticles functionalized with aminethiols, polycationic electrolytes such as poly(ethyleneimine), etc.

On the other side and as discussed before, the chemical reactivity of CNT side walls increases with the tube curvature (e.g. decrease of the tube diameter), due to the increase of the pyramidalization angle and greater strain energy per atom [34, 35]. Such pyramidalization of the CNT atoms causes the exohedral lobes of the orbitals to be larger than their endohedral counterparts. The reactivity of the surface is thus enhanced by the pronounced exposure of the hybrid orbitals from the exterior which favours the orbital overlap with incoming reactants [35]. Accordingly, many reactions can be performed on the sidewalls of the CNTs, such as halogenation, hydrogenation, radical, electrophilic and nucleophilic additions, etc. [10, 34, 36, 162, 170, 172]. Exhaustive explored examples are the nitrene cycloaddition, the 1,3-dipolar cycloaddition reaction (with azomethinylides), radical additions using diazonium salts or radical addition of aromatic/phenyl primary amines. The aryl diazonium reduction can be performed

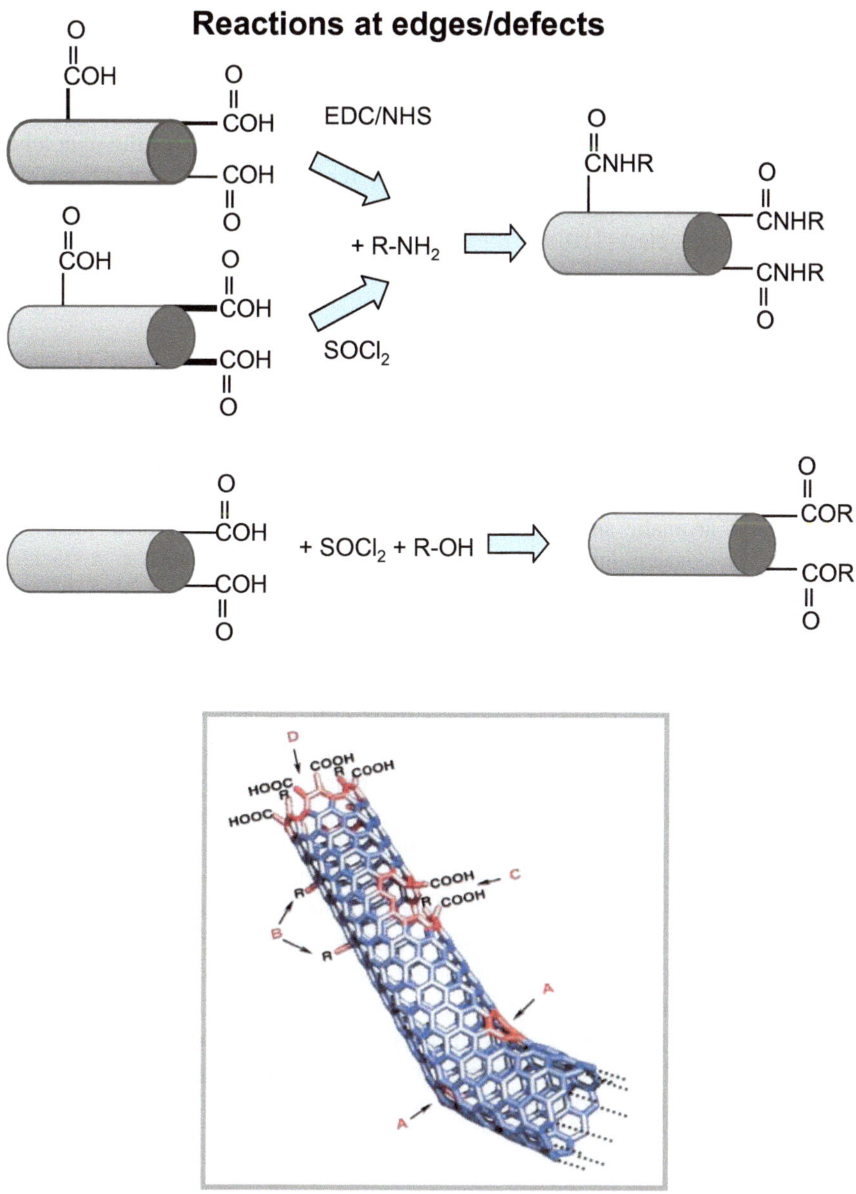

Fig. 1.25 Scheme showing common covalent reactions at the CNT edge/defects and at the CNT side walls: carbodiimide chemistry. The figure also shows the typical defects in a SWNT: **a** five- or seven-membered rings in the C network, instead of the normal six-membered ring, leads to a bend in the tube; **b** sp3-hybridized defects (R = H and OH); **c** carbon network damaged by oxidative conditions, which leaves a hole lined with –COOH groups; and **d** open end of the SWNT, terminated with oxygenated species such as –COOH groups. The last part figure has been reprinted with permission from Ref. [162]

by electrochemical means by forming a phenyl radical (by the extrusion of N_2) which couples to a double bond [170]. Similarly, electrochemical oxidation of aromatic or aliphatic primary amines yields an amine radical which can be added to the double bond on the carbon surface. The direct covalent attachment of functional moieties to the sidewalls strongly enhances the solubility of the nanotubes in solvents and also tailors the CNT surface for the different applications (for instance, in medical chemistry, solar energy conversion, selective recognition of chemical species, etc.) [170].

Non-covalent Modification

Covalent modification yields a very stable and effective derivatization. However, some drawbacks can emerge when pursuing applications especially with SWNTs. A high density of reactive sites or covalently functionalized points can lead to a loss of the CNT conjugation network with the consequent degradation of the CNT mechanical and electronic properties.

The large aromatic and hydrophobic character of CNTs make them ideal surfaces for non-covalent interaction with molecules via Van der Waals, π–π stacking or hydrophobic forces [36, 170]. There are many examples in literature of the non-covalent binding of amphiphilic agents on CNTs which have promoted CNT disentanglement and made them water soluble and biocompatible. Examples of such agents are some organic ionic surfactants such as sodium dodecyl sulfate or dodecylbenzene sulfonate. Other examples include polyaromatic compounds such as pyrenes with hydrophilic functional groups, proteins or polymers (polyethyleneglycol, tween 20, Triton X-100, polyvinylpyrrolidine, polystyrene sulfonate, DNA, RNA, polysaccharides, etc.) which coat or wrap the nanotube in such a way that the more hydrophobic part interacts with the CNT surface whereas the hydrophilic region faces the aqueous medium. An illustration is depicted in Fig. 1.26. Some of these polymers have also been used as blocking agents of non-specific binding of proteins in the development of biosensors (PEG, tween 20, Triton X-100).

A variety of these physisorbed molecules were used as anchor points for covalent immobilization in a so called hybrid approach [171]. That is the case of bifunctional pyrenes (e.g. 1-pyrenebutanoic acid succinimidyl ester, pyrene maleimide) or polymers with terminal amine/carboxylic groups which have been used for covalently immobilization of proteins, functionalized oligonucleotides, etc. (Fig. 1.26) [170, 172].

1.3.2.2 The Biosensing Event

As stated, in this section we will describe the biosensing events that can take place on the different CNT platforms together with the mechanism of action of the biological recognition elements. Specifically, we will discuss about Enzymatic and redox protein biosensors, CNT/DNA and genosensors, Immunosensors and CNT-Aptasensors.

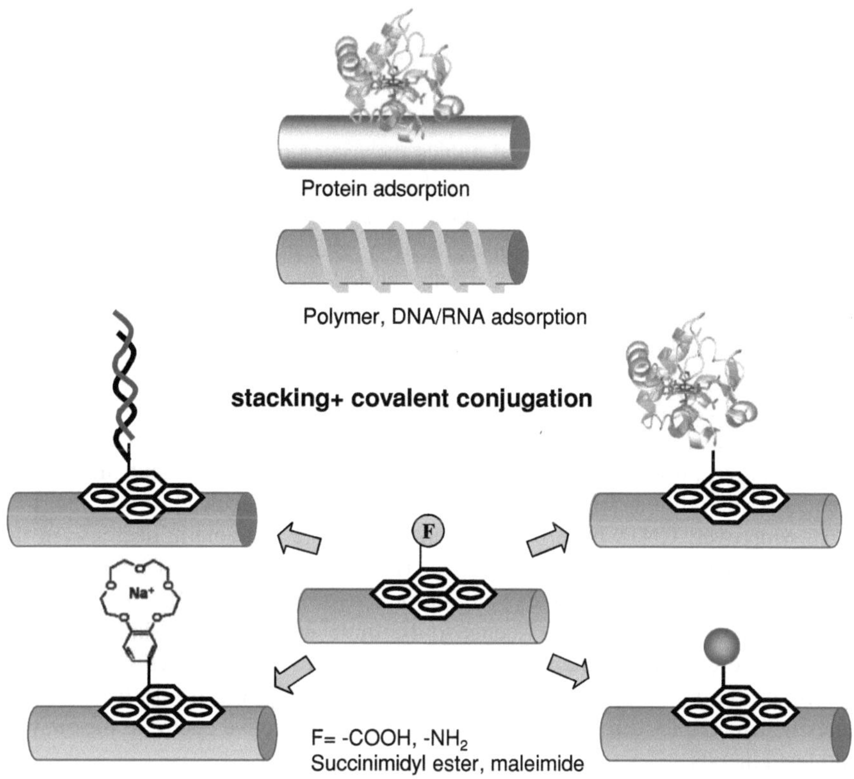

Fig. 1.26 Schematic representation of the typical non-covalent CNT functionalization and the hybrid approach by using pyrene linkers

Enzymatic and Redox Protein Biosensors

Electron transfer in biological systems is a very interesting topic not only because fundamental studies of such processes can help to unravel basic mechanisms in nature machinery, but also because such processes can be mimicked and exploited to develop sensors, bioreactors or novel fuel cells. In the electroanalytical field, there is an important interest in studying electron transfer between a redox biomolecule and an electrode for amperometric (bio)sensors. Such sensors are based on the ability of an enzyme to transduce the turnover with its corresponding substrate into a detectable, reliable and quantifiable current. One of the main goals in electroanalysis and mainly in enzymatic sensors is to achieve direct electron transfer between the protein redox centre and the transducer, keeping at the same time its bioactivity [40, 42, 51, 124, 172–174]. Thus, the electrode itself can replace physiological partners or natural electron acceptors, providing the driving force to energize the reaction and a sensor to measure the response. For instance, in the case of glucose oxidase (GOx), which catalyses the reaction:

glucose + O_2 → glucolactone + H_2O_2, with oxygen acting as its natural electron acceptor to regenerate the enzyme, the electrode can take the role of its natural partner oxygen.

A direct electron transfer warrants an efficient transduction of the enzymatic recognition of a target analyte and avoids the use of mediators or low-molecular weight relays [172]. However, such goal is rather difficult to achieve since proteins, when adsorbed on electrode surfaces, undergo denaturation with the consequent loss of their electrochemical activity and bioactivity. Moreover, some enzymes contain the redox active centre buried deep inside the protein. That makes generally difficult to obtain direct electron exchange between the protein and electrode surfaces so that mediators are needed to obtain an electrochemical response. Therefore, electrode materials with suitable physicochemical properties and enzyme immobilization methods are important for obtaining the direct electron transfer and preserving their bioactivities. CNTs seem to be an ideal material to immobilize proteins and promote direct electron transfer without the need of mediators and without losing the bioactivity of the biomolecules [40, 42, 124, 172, 173]. Moreover, CNTs have been shown to promote a dramatic decrease in the overpotential of some important species (H_2O_2, NADH) involved in electrocatalytic enzymatic reactions such as oxidases or dehydrogenases [40].

The use of aligned CNTs seems to be quite advantageous since the more electroactive ends of the nanotubes are readily accessible to species in solution. Moreover, the rigidity of the tubes allows them to be plugged into the biomolecules which enables direct electrical connection to the active centre of the enzyme and fast kinetics.

Direct electron transfer has also been achieved with many metalloproteins such as cytochrome C, horseradish peroxidase, microperoxidase (MP-11), myoglobin, hemoglobin, catalase, azurin, etc. immobilized on different CNT modified electrodes [44, 53, 134, 175–183]. The electrochemical performance of typical metalloproteins on different CNT electrodes will be discussed in this thesis work taking as an example the small water soluble myoglobin (Mb) and catalase (Cat).

Myoglobin and Catalase: Iron Based Proteins

Myoglobin is a small oxygen-binding protein of muscle cells, which stores oxygen and facilitates oxygen diffusion in rapidly contracting muscle tissue. It has a molecular weight of 16,700 Da and is comprised of a single polypeptide chain of 153 aminoacid residues of known sequence and a single iron protoporphyrin IX as the prosthetic group allocated in a hydrophobic pocket [184, 185]. At the centre of protoporphyrin, the iron atom is bonded to nitrogen atoms from four pyrrole rings. The protein contains a proximal histidine group attached directly to the iron centre, and a distal histidine group on the opposite face, not bonded to the iron. The distal side of the heme is either vacant or occupied by a water molecule which is easily exchanged by other ligand such as oxygen or hydrogen peroxide [173], Fig. 1.27. The distal histidine stabilizes a water-ligand to ferric iron and in the ferrous state this conformation stabilizes bound oxygen as the sixth ligand and

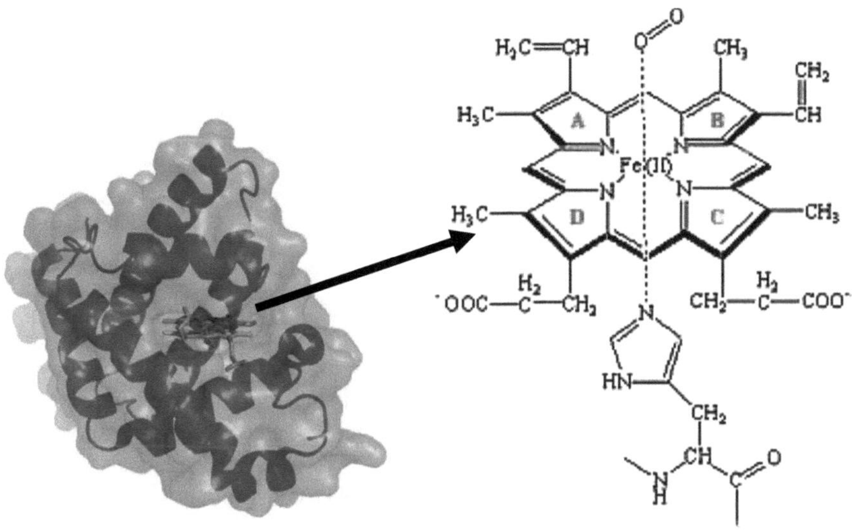

Fig. 1.27 Myoglobin molecule contains one heme prosthetic group inserted into a hydrophobic cleft in the protein. Representation of heme prosthetic group (biologically active group in Mb), its interaction with the polar proximal histidine and the oxygen ligand at the 6th axial site

suppresses the irreversible autooxidation of the heme group [173]. Thus, the distal histidine acts as a gate by controlling the access of ligand molecules to the distal pocket. The outside of the protein has both polar and apolar residues with an isoelectric point[11] of 6.8. Besides its important biological task, Mb attracts attention due to its hemin core capable of redox activity, which is unusual for a protein not naturally involved in electron transfer.

On the other hand, catalase is a large common antioxidative enzyme present in the peroxisomes of nearly all aerobic cells, serves to protect the cell from the toxic effects of hydrogen peroxide (H_2O_2) by catalyzing its decomposition into molecular oxygen and water without the production of free radicals [186–190]. All catalases are tetrameric, each subunit (molecular weight ~ 60,000 Da) consists of 506 aminoacid polypeptide chain with hemin as a prosthetic group which is embedded in the middle of each monomer. Catalase is one of the most efficient enzymes known, resulting in reaction rates approaching the diffusion-controlled limit. The optimum pH for human catalase is approximately 7.

As mentioned above, both proteins have the active redox centre embedded deep in the protein which makes difficult the direct electron transfer in conventional electrodes.

However, the transfer can be enhanced with a proper immobilization of the protein on adequate substrates. Additionally, the redox iron centre of both proteins is extremely sensitive to oxygen and peroxide and electrocatalytical activity can be

---

[11] Isoelectric point is the pH at which the number of positive charges and the number of negative charges of a compound (e.g. protein) are equal.

observed in presence of such analytes by monitoring the electron transfer between the electrode and the protein iron group. The possibility of enhancing the electrochemical response of this metalloproteins together with its high sensitivity to oxygen and H_2O_2 gives rise to a strong interest, specially, in the area of the electroanalysis, for the development of efficient electrochemical recognition (bio)sensors for clinical diagnosis or for environmental assays.

CNT/DNA and Genosensors

The modification of CNT with nucleic acids constitutes a very promising area for the development of devices for DNA-detection, gene therapy, drug discovery and delivery, etc. A genosensor consists in a substrate modified with specific oligonucleotides (probe DNA) that can detect targeted complementary DNA sequences (target DNA). The application of CNTs in electrochemical DNA biosensors is quite promising for many reasons: CNTs constitute novel platforms for DNA immobilization with important electrochemical transducing properties and high specific surface area which can increase not only the attached DNA amount but also concentrate a great number of enzymes or electroactive nanoparticles to amplify DNA hybridization [40, 173].

In order to prepare electroanalytical devices based on hybridization recognition, the first critical issue to face is related with the immobilization procedures on the surface of the transducer. It is important to guarantee a good orientation of the bases of the DNA probe on the CNT interface in order to get efficiency in the hybridization with the target DNA. The more basic approaches for DNA immobilization on CNTs are the physisorbed (multisite attachment) or the covalent anchoring (single point linkage). In the latter case, typical schemes are provided by the conjugation of DNA terminal functional groups with functional groups in the CNT surface via the carbodiimide chemistry, streptavidin/biotin interactions, etc.

In general, DNA is physisorbed on CNTs by wrapping the tube, with the bases (apolar region of the molecule) in close contact with the hydrophobic walls of CNT whereas the phosphate backbone faces the solution [170, 191]. Such orientation of the adsorbed DNA and the consequent reduced flexibility can decrease the efficiency of the hybridization event, indicating that for a more appropriate biofunctionalization, hydrophobic interactions need to be minimized. Therefore, when genosensors based on physical adsorption of oligonucleotides are considered, chemical treatment of the CNT surface (by promoting hydrophilic interactions) should be performed to increase the efficiency of the hybridization detection. On the other hand, single-point attachment of the DNA turns out to be a more flexible configuration which enhances hybridization kinetics.

Many of the electrochemical DNA-detection strategies that can be collected in the literature range from the use of labelling target DNA to label-free approach. Among the later, the direct electrochemical detection of DNA methodologies such as the oxidation of the DNA bases (mainly guanine) or the use of highly positively charged o-negative redox reporters as indicators of the hybridization process are of considerable interest, though the indirect detection methods of DNA hybridization by using labels (i.e. covalently linked fluorescent tag DNA target) are very popular and in some cases mandatory [192].

Label-free electrical detection protocols based on DNA oxidation are attractive since they greatly simplify DNA hybridization assays as they offer an instantaneous detection of the duplex. However, such kind of detection schemes can suffer from some drawbacks [193]. DNA bases exhibit high oxidation potential and consequently higher background currents which can mask the electrochemical signal [194, 195]. In some cases, it is also necessary to manipulate the DNA probe by replacing guanine by inosine in order to avoid interferences of the guanine oxidation from the probe when monitoring the DNA target [192]. Other additional issues that can perturb the performance of the biorecognition event can arise from the functionalization process. For direct oxidation, a DNA multisite attachment provided by physisorption is required to guarantee close proximity between the CNT and the DNA for electron exchange. As mentioned, such configuration reduces flexibility of DNA for the hybridization process. A covalent single point attachment could overcome such difficulties but then base oxidation mediators (such as ruthenium complexes) should be used for achieving a quantifiable signal since the bases can not remain in close contact with the electrode surface.

Labelled DNA detection schemes rely on the chemical labelling of target DNA sequences with redox active molecules, enzymes, nanoparticles, etc. [117, 192, 196–199]. In the case of enzymatic labels bound to the DNA target, the enzyme triggers the catalysis of a redox active reaction under hybridization. For instance, by using a horseradish peroxidase-labelled DNA target, H_2O_2 peroxide is produced which can be detected amperometrically, so the redox reaction of H_2O_2 is an indication of the hybridization process. A variation on such approach involves a three-component "sandwich" assay, in which the redox label is attached to a synthetic sequence specifically designed to bind an overhang portion of the probe and target DNA. All these approaches need an extra chemical labelling step either in the target DNA or in the synthetic oligonucleotide which makes the process more expensive and effortful [193]. One practical alternative to DNA base oxidation detection or chemical labelling schemes is to use redox active reporter molecules that intrinsically associate with the double helix in a non-covalent manner. These reporter molecules can interact with the DNA either by electrostatic interactions or by intercalation in the double strand DNA. A simple example of the former molecules is the highly positively charged $Ru(NH_3)_6^{3+}$ which can strongly interact with the negatively phosphate backbone. Such approach with non-covalent and reversible redox reporters is very sensitive, simple and does not damage the sample which allows further use.

Immunosensors

An immunosensor is a device comprising an antigen or antibody species coupled to a signal transducer, which detects the binding of the complementary species. An extensive range of analytes can be detected and measured by immunosensors, e.g.

medical diagnostic markers such as hormones (steroids and pituitary hormones), drugs (therapeutic and abused), bacteria and environmental pollutants such as pesticides.

An indirect immunosensor uses a separate labelled species that are detected after binding by, e.g. fluorescence or luminescence (e.g. a heterogeneous immunoassay). A direct device detects the binding by a change in potential difference, current, resistance, mass, heat, or optical properties (e.g. a homogeneous immunoassay). Although indirect sensors may encounter fewer problems due to non-specific binding effects, the direct sensors are capable of real-time monitoring of the antigen–antibody reaction. A wide range of molecules can be detected with detection limits ranging between 10^{-9} and 10^{-13} mol L^{-1}.

Although it has not been dealt in this thesis work, CNTs have also promising applications in the immunosensing field and many examples of such sensor devices can be found in literature. For example, immunosensors with an electrochemiluminescence read-out signal have been designed using CNTs as support of the immunorecognition systems [200].

As an alternative to immunosensors, we have developed CNT-aptasensors which confer some advantages when we use specific detection schemes, as for instance, based on field-effect transistors, but this will be discussed in the following section.

CNT-Aptasensors

Advances in nucleic acid research led to the identification of specific oligonucleic acid sequences, which bind to wide array of non-nucleic acid target molecules like proteins, with high affinity and specificity. Such nucleic acid sequences are termed as aptamers [201]. The word aptamer derives from aptus that means 'to fit' [202]. These are RNA or DNA molecules (ca. 30–100 nucleotides) that recognize specific ligands and that are obtained in vitro by 'selection evolution of ligands with exponential enrichment', SELEX process [203].

The SELEX process (Fig. 1.28) starts by generating a large library of randomized RNA or DNA sequences. Usually, this library contains 10^{14}–10^{15} different RNA/DNA species that fold into different structures depending on their particular sequence. The library is incubated with the target protein of interest, and those RNAs/DNAs present in the library that bind the protein are separated from those that do not. The retained RNAs/DNAs are then amplified by RT-PCR and transcribed in vitro to generate a pool of RNAs/DNAs that have been enriched for those that bind the target of interest. This selection and amplification process is repeated (usually 8–12 rounds) until the RNA/DNA ligands with the highest affinity for the target protein are isolated. The winning aptamers are then cloned and sequenced.

Aptamer/protein binding results in a highly specific interaction, which has the ability to discriminate between related proteins that share common sets of structural domains [204–207].

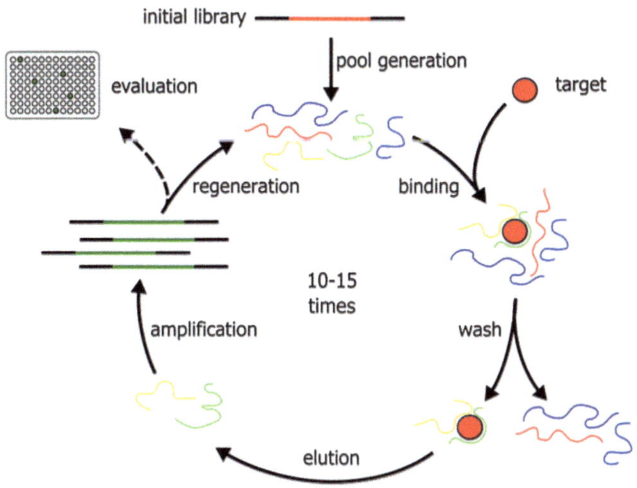

Fig. 1.28 Scheme of systematic evolution of ligands by exponential enrichment (SELEX). A library of DNA or RNA molecules is incubated with the protein target, and those that bind to it are separated from the rest. The sequences with affinity for the target are subsequently amplified to generate a pool of molecules that bind to the protein of interest

Aptamers are often called 'synthetic antibodies' because they can mimic antibodies in a number of applications and bind their targets with affinities and specificities that can be comparable to those of antibodies: the equilibrium dissociation constants of aptamers to targets are usually in the range of picomolar (pM) to micromolar (μM), similar to those of antibodies for antigens [208, 209]. However, aptamers present some advantages. Table 1.2 summarizes them [210]. Aptamers can be used not only for the molecular recognition in immunoassay-like tests, but also as sensing elements in biosensors. Biosensors based on aptamers as biorecognition elements have been coined 'aptasensors' [211]. The grey-coloured rows detail special advantages of aptamers over antibodies as recognition elements.

In the past few years, integration of functional aptamers into nanomaterials has become a new interdisciplinary field that aims at providing new hybrid sensing systems (sensors) for specific and sensitive molecular recognition [212, 213]. This novel integration has yielded various types of sensors for selective and sensitive detection of a wide range of analytes such as adenosine, cocaine, mercuric ion, thrombin, etc.

In this thesis work, the preparation, characterization and applications of CNTs that are conjugated with aptamers will be discussed, as well as the suitability of using different configurations of CNT (randomly dispersed on GC, VACNT, FET, etc.). It is worth highlighting that in terms of FET technology, aptamers provide a preferable choice over antibodies that will be discussed in detail later on. The study will be focused on aptamer nanosensors for the biosensing of proteins like thrombin and lysozyme.

Table 1.2 Advantages of aptamers over antibodies

	Aptamers	Antibodies
Binding affinity	nM-pM	nM-pM
Specificity	High	High
Production	Chemical process carried out in vitro and can therefore target any protein	Requires in vivo biological system
Target range	Wide: ions, small organic molecules, proteins, whole cells, etc.	Narrow: only immunogenic compounds
Batch to batch variation	Uniform activity regardless of batch synthesis	Activity of antibodies vary from batch to batch
Chemical modification	Easy and straightforward	Limited, can cause loss in affinity
Target site	Investigator determines target site of protein	Immune system determines target site of protein
Thermal denaturation	Return to original conformation after temperature treatment	Temperature sensitive and undergo irreversible denaturation
Shelf-life	Stable to long term storage and can be transported at ambient temperature	Limited
Cross-reactive compounds isolation	Can be isolated utilizing toggle strategy to facilitate pre-clinical studies	No method

Antibody generation, particularly for use in biosensors, has several drawbacks that are addressed by aptamers

Thrombin and Lysozyme

Thrombin

Thrombin is a serine protease that plays many roles in the coagulation cascade, it converts soluble fibrinogen into insoluble strands of fibrin, as well as catalyzes many other coagulation-related reactions [214]. Thrombin is a two chain enzyme composed of an NH_2-terminal "A" chain composed of 36 residues (Mr = 6,000 Da) and is non-essential for proteolytic activities and a COOH-terminal "B" chain composed of 256 amino acids (Mr = 31,000 Da) which remain covalently associated through a single disulfide bond. Human thrombin is 13 amino acids shorter than the bovine thrombin due to a thrombin cleavage site on the human protein that is not present in the bovine protein. Its isoelectric point is 7.0–7.6 [215]. Thrombin is usually considered as an important target when searching for anti-coagulants and antithrombotics to interfere in the blood coagulation [216]. Moreover, thrombin is considered as an useful tumour marker in the diagnosis of pulmonary metastasis [217]. Disorders in blood clotting are tightly linked to many serious health issues, including heart attacks and strokes [218]. Therefore, thrombin is typically the target in anticoagulation therapy for these diseases [219].

Lysozyme

Lysozyme, also called muramidase or peptidoglycan N-acetylmuramoyl-hydrolase, is an ubiquitous enzyme widely distributed in diverse organisms such as bacteria, bacteriophages, fungi, plants and animals [220]. It catalyzes in vivo the hydrolysis of the β(1–4) glycosidic linkage between N-acetylmuramic acid and N-acetylglucosamine alternating sugar residues in the bacterial peptidoglycan and causes bacterial cell lysis. Its primary sequence contains 129 amino acid residues. The molecular weight is 14,351 Da and its isoelectric point is 11.0 [221, 222]. Lysozyme's relatively small size and simplicity makes it an excellent model analyte for novel methods in protein detection. This enzyme works as a natural inner body antibiotic often termed "body's own antibiotic" because it possesses lytic activity against the polysaccharide wall of bacteria. Moreover, it has been discovered recently that antibodies against variants of lysozyme are present in patients with rheumatoid arthritis [223] which highlights the potential of lysozyme as a clinical index of these diseases.

1.3.2.3 Special Issues for the Biorecognition Process

Selectivity is especially important in real-world samples where the target concentration can be much less than the concentration of non-target biomolecules. Thus there are some issues that need to be taken into consideration for an efficient biorecognition event.

Blocking Agents

A trade-off exists between selectivity requirements and sample preparation complexity for most real-world applications. Most authors opine that obtaining adequate selectivity in complex real-world samples is the most daunting challenge to the field of biosensors in general [224]. A closely related concept is non-specific binding (NSB), in which non target biomolecules stick to the probe layer, preventing target binding or causing a false positive signal. To alleviate this problem, one should pre-expose the sensor modified with probe molecules to a solution containing specific blocking agents: molecules that cover the sensor gaps, for instance in our case the CNT walls left unprotected by the probe, thus preventing the non-specific adsorption of interferents when exposing to the specific target analytes.

However, there is no previous study which deals systematically in depth the problem of blocking the non-specific adsorption on the CNTs. Experimental applications are limited to blocking agents widely used to prevent protein adsorption on other surfaces, such as bovine serum albumin (BSA) or other antifouling agents such as polyethylene glycol (PEG) [225, 226], polyethylene glycol methacrylate [227–229] and surfactants like Tween 20 [226, 230, 231] that can also be deposited on areas surrounding the sensor to prevent target depletion via non-specific

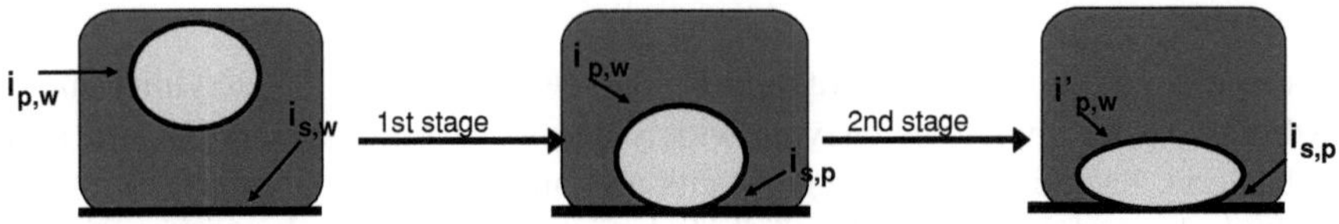

Fig. 1.29 Illustration of the proteins adsorption mechanism in two stages, where: *p* protein, *i* interface, *s* surface, *w* water. After the second stage, conformational changes can modify the interface with water resulting in i'_{pw}

binding [232, 233]. These agents have been widely used for biomedical applications and biosensors.

Taking as an example the non-specific adsorption of proteins on CNT, the major problem comes from the CNT hydrophobic surface and the fact that proteins are capable of conformational changes that increase their interactions (hydrophobic interactions) with surfaces [234].

Specifically, there are enthalpic and entropic contributions in the protein adsorption processes on surfaces. Enthalpic contributions include Van der Waals, hydrophobic and electrostatic (between the surface and the oppositely charged protein) interactions. Entropic contributions are based on mechanisms involving the release of counterions and/or water molecules that form the hydration shell, as well as conformational changes [234].

The protein adsorption mechanism considers two stages, being the first one more decisive (Fig. 1.29). It is based on two interfaces, one between the surface and the water and the other between protein and water. The first step consists in the formation of a new interface between the surface and the protein, with the consequent displacement of water. The second stage considers the reorganization of the protein on the surface.

It is possible to develop a strategy to block non-specific adsorption on CNT hydrophobic surface if the first stage is made difficult. One way to accomplish that is by modifying the CNTs surface with blocking agents that may confer a stable hydrophilic interface that hinders the displacement of water molecules from the CNT-protein interface.

From studies of self-assembled monolayers with terminal functional groups ethylene glycol (EG), it was demonstrated the capacity of these groups to stabilize water molecules by hydrogen bonds, and consequently, the generation of a solvation layer at the interface which energetically suppress adsorption of proteins on the EG modified surface [235]. Although the molecular mechanisms underlying protein resistance of ethylene glycol moieties have not yet been fully identified, it is believed that apart from the stabilization of hydration shells there is also a brush-induced “steric” repulsion that is thought to prevent direct contact between proteins and the underlying surface.

Apparently, studies that have tested different blocking agents with EG groups (PEG, Triton X-100, Tween 20, etc.) conclude that the blocking agent that has less number of EG units is the one which confers less resistance to protein adsorption.

However, this view is far from giving a complete explanation of the process; layout and density of the blocking agents on CNT should be taken into account and therefore, one should use more specific techniques that give information for instance on coating degrees, orientation and thickness of the agents chains such as surface plasmon resonance, grazing incidence X-ray diffraction, theoretical studies based on molecular dynamics simulations, etc.

The use of blocking agents has not been systematically studied, but several approaches have been found to work in specific situations. For instance, washing the sensor surface before readout can sometimes improve selectivity by washing away non-specifically adsorbed molecules while leaving the target intact, but this in an endpoint measurement and not real-time approach.

1.4 Characterizing the Biosensor Devices

In this part we will provide an overview of the fundamentals of the main techniques that have allowed characterizing the biosensor devices and following the sensing events. Specifically, we will describe the different electrochemical and electronic methods of detection used and surface characterization techniques.

1.4.1 Methods of Detection: Electrochemical and Electronic Biorecognition Processes

Although optical detection techniques are perhaps the most prevalent in biology and life sciences, electrochemical or electronic detection techniques have also been used in biosensors/biochips due to their great sensitivity, high specificity, and low cost. These techniques can be amenable to portability and miniaturization, when compared to optical detection techniques.

In the next lines, the electrochemical and electronic techniques used to characterize the biosensor devices of this thesis work will be described. These techniques include voltammetric techniques (cyclic voltammetry and differential pulse voltammetry), chronocoulometry, electrochemical impedance spectroscopy and electronic detection based on electric field.

1.4.1.1 Voltammetric and Chronocoulometric Measurements

Voltammetric techniques involve the application of a potential (E) to an electrode and the monitoring of the resulting current (i) flowing through the electrochemical cell. In many cases, the applied potential is varied over a period of time (t) while monitoring the current. Among all the common forms of voltammetry, we have focused in cyclic voltammetry and differential pulse voltammetry.

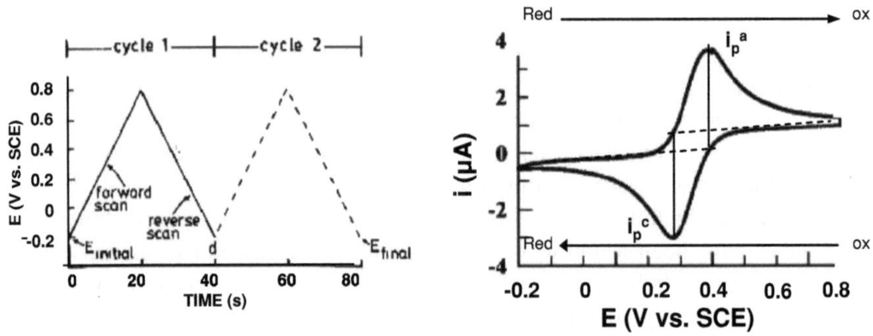

Fig. 1.30 **a** Typical excitation signal for cyclic voltammetry, triangular potential waveform with switching potentials at −0.2 and 0.8 V versus SCE; **b** cyclic voltammogram scheme

Cyclic voltammetry (CV) is perhaps the most versatile electroanalytical technique for the study of electroactive species [236]; it can provide qualitative information about the number of oxidation states and their stability, about mechanisms of electrochemical reactions, as well as their rate of heterogeneous electron transfer. It can also be used in combination with simulation software to calculate rates of homogeneous and heterogeneous reactions.

Cyclic voltammetry is often the first experiment performed in an electrochemical study of a compound, a biological material, or an electrode surface. For instance, CV allows following biorecognition events with electron transfer exchange. The effectiveness of CV results from its capability for rapidly observing the redox behaviour over a wide potential range. The resulting voltammogram is analogous to a conventional spectrum in that it conveys information as a function of an energy scan.

Specifically, CV consists of cycling the potential of an electrode, which is immersed in an unstirred solution, and measuring the resulting current. The potential of this working electrode is controlled versus a reference electrode such as a saturated calomel electrode (SCE) or a silver/silver chloride electrode (Ag/AgCl). The controlling potential which is applied across these two electrodes can be considered an excitation signal (for CV is a linear potential scan with a triangular waveform) as shown in Fig. 1.30.

The important parameters in a cyclic voltammogram are the peak potentials (E_p^c, E_p^a) and peak currents (i_p^c, i_p^a) of the cathodic and anodic peaks, respectively. If the electron transfer process is fast compared with other processes (such as diffusion), the reaction is said to be electrochemically reversible, and the peak separation is

$$\Delta E_p = E_{pa} - E_{pc} = 2.303 \cdot RT/nF.$$

Thus, for a reversible redox reaction at 25 °C with n electrons ΔE_p should be $0.0592/n$ V.

The formal reduction potential (E^o) for a reversible couple is given by:

$$E^o = \frac{E_{pc} + E_{pa}}{2}$$

For a reversible reaction, the concentration is related to peak current by the Randles–Sevcik expression (at 25 °C):

$$i_p = 2.686 \times 10^5 n^{\frac{3}{2}} A c_0 D^{\frac{1}{2}} \upsilon^{\frac{1}{2}}$$

where i_p is the peak current in amps, A is the electrode area (cm^2), D is the diffusion coefficient ($cm^2\ s^{-1}$), c_0 is the concentration in mol cm^{-3}, n the number of electrons and v is the scan rate in V s^{-1}.

Accordingly, the current is directly proportional to concentration and increases with the square root of the scan rate. A typical cyclic voltammogram is shown in Fig. 1.30b.

For the cases that the electron transfer processes are "slow" (relative to the voltage scan rate) the reactions are referred to as quasi-reversible or irreversible electron transfer reactions. In this situation the voltage applied will not result in the generation of the concentrations at the electrode surface predicted by the Nernst equation. This happens because the kinetics of the reactions is 'slow' and the equilibrium is not established rapidly (in comparison to the voltage scan rate). In this situation, the overall form of the voltammogram recorded is similar to that above, but unlike the reversible reaction, now the position of the current maximum shifts depending upon the reduction rate constant (and also the voltage scan rate). This occurs because the current takes more time to respond to the applied voltage than in the reversible case.

In the irreversible processes, the individual peaks are widely separated and the peak current is given by:

$$i_p = 2.99 \times 10^5 (\alpha n_a)^{\frac{1}{2}} n A C_0 D^{\frac{1}{2}} \upsilon^{\frac{1}{2}}$$

with α being the transfer coefficient and n_a is the number of electrons involved in the charge transfer step. The current is still proportional to the bulk concentrations but will be lower in height depending on the value of α. As before, the peak current increases linearly with $\nu^{1/2}$.

For quasi-reversible systems, the shape of the peak and the peak parameters are functions of α and a parameter Λ, the latter one defined as:

$$\Lambda = \frac{k}{(D\upsilon nF/RT)^{\frac{1}{2}}}$$

where k is the heterogeneous electron transfer rate constant.

As Λ increases, the process approaches the reversible case. For small values of Λ, the system exhibits a more irreversible behaviour. Then, the peak current is given by:

$$i_p = i_p(rev) K(\alpha, \Lambda)$$

where $i_p(rev)$ is the reversible value of i_p and K is a function of Λ and α. In this case i_p is not proportional to $\nu^{1/2}$.

Differential pulse voltammetry (DPV) is an important electroanalytical tool which provides increased sensitivity and more efficient differentiation and resolution of different species compared to CV.

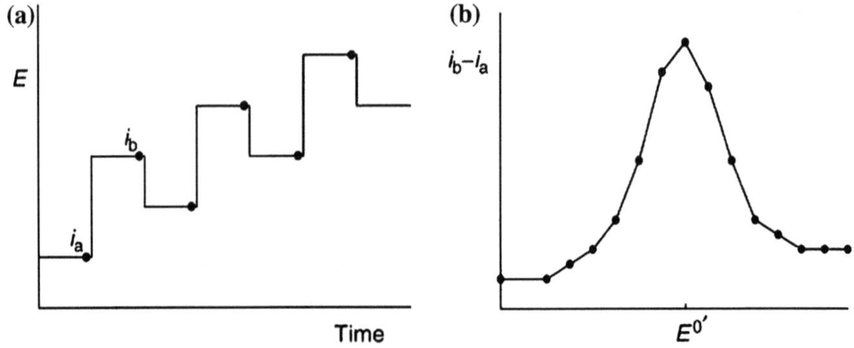

Fig. 1.31 **a** Diagram of the potential waveform used in DPV; **b** differential pulse voltammogram

This technique uses a series of fixed potential pulses of small amplitude (10–100 mV) which are superimposed on a slowly changing base potential as shown in Fig. 1.31.

The current is sampled immediately before each potential pulse is applied (i_a) and immediately before each pulse ends (i_b), and it is the difference between these current responses ($i_b - i_a$), that is used to produce the voltammogram. By sampling the current just before the potential is changed, the effect of the charging current can be decreased.

Chronocoulometry (CC) is one of the classical electrochemical techniques frequently used in electroanalytical chemistry. As its name implies, CC is the measurement of charge (coulombs) as a function of time (chrono). Applications of this technique include measurement of electrode surface area, diffusion coefficients, concentration, kinetics of both heterogeneous electron transfer reactions and chemical reactions coupled to electron transfer, adsorption, and the effective time window of an electrochemical cell. In our particular case, chronocoulometry measurements can help to estimate the accumulated cationic redox marker at the oligonucleotide CNT modified surface as indicative of the hybridization process [237].

In this method a pulse of potential is applied and the response in current is monitored as a function of the time. From the integration of the current profiles versus time, one can obtain the charge (Q_{TOT}). The charge contains information of the double layer charging (at very short times), the charge of the species adsorbed and the charge of the species diffusing at the electrode (longer times) according to the following relation :

$$Q_{TOT} = Q_{dl} + Q_{ads} + Q_{diff} = Q_{dl} + nFA\Gamma_o + \frac{2nFAD_o^{1/2}C_o}{\pi^{1/2}}t^{1/2}$$

The total charge (Q_{TOT}) measured in response to the potential step thus, comes from three sources:

(a) Charge of the double layer (Q_{dl});
(b) Charge of the adsorbed species (Q_{ads});
(c) Charge of the species diffusing at the electrode (Q_{diff}).

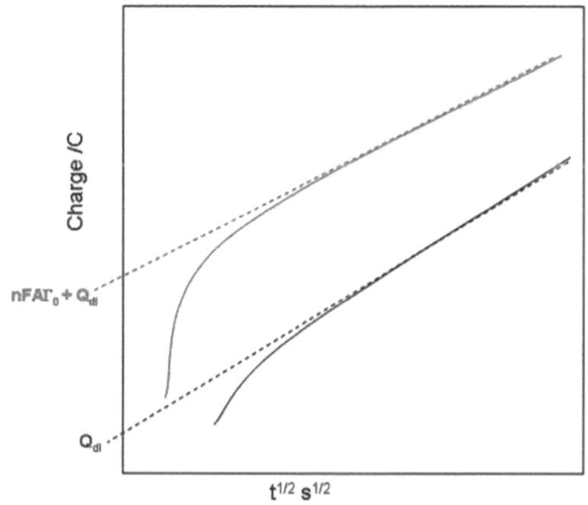

Fig. 1.32 Illustration of the procedure in order to get the amount of electroactive species adsorbed on an electrode, Γo_0 The *red curve* represents the chronocoulometric measurements for an electrode immersed in an electrolyte that contains low concentration of electroactive species that can be adsorbed on the electrode surface. The *black curve* is the chrono measurement in presence of the background electrolyte but in absence of the electroactive species (Color figure online)

F is the Faraday constant (96,500 C), A is the electrode area, n the number of electrons and C_0 is the concentration of the redox species at the bulk electrolyte, D is the diffusion coefficient of the species and Γ_0 is the amount of adsorbed species at the interface of the electrode.

One of the important applications of CC is the estimation the amount of electroactive species adsorbed on the surface, for which one has first to subtract the double layer charging [237]. To estimate it, chronocoulometric measurements in the background electrolyte (without the presence of the electroactive species that can undergo adsorption on the electrode) are performed. By extrapolation from the linear region of the curve of charge as a function of $t^{1/2}$, one can obtain the ordinate at the origin which corresponds to the double layer charge (Q_{dl}). Then the chrono measurements are performed in presence of the redox markers that can adsorb on the surface. Again, the ordinate of the origin is estimated which comprises the charging of the double layer plus the charge of the adsorbed amount (Γ_0). Therefore, from both measurements one can get the adsorbed amount by Q_{dl} subtraction as shown in Fig. 1.32.

1.4.1.2 Electrochemical Impedance Spectroscopy

Electrochemical Impedance spectroscopy (EIS) is a very versatile tool widely used in different fields (corrosion [238, 239], semiconductor electrodes [240, 241], polymers and coatings [242, 243], batteries and fuel cells [244–246], electrode kinetics and mechanisms [247, 248], biomedical and biological systems [224, 249], solid-state systems [250], etc.). Since EIS turns out to be very sensitive to probe the interfacial properties of a modified electrode, it has become an attractive tool to monitor biorecognition events at the electrode surface [251–253].

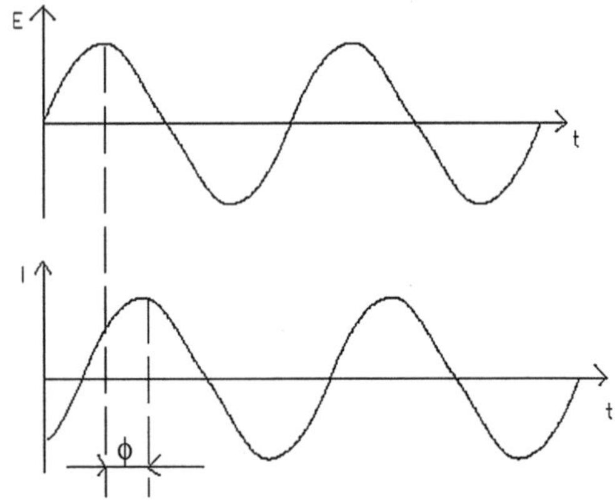

Fig. 1.33 AC excitation signal applied and sinusoidal current response in the system under study

Specifically, EIS allows characterizing the double layer interface at the electrode, the physicochemical processes of a widely differing time constants, sampling electron transfer at high frequencies and mass transfer at low frequencies. The processes occurring in the electrochemical cell can be modelled by using combination of resistors and capacitors (equivalent circuits) [254]. The experimental spectra can be fitted with the use of equivalent circuits and obtain the values of electrical parameters such as resistances, capacitances, etc., which characterize the electrochemical interface.

Voltammetry or amperometry involve measuring the direct current (DC) current at an electrode as a function of applied electrode-solution voltage. In contrast, electrochemical impedance (Z) is based on the collection of an alternating current (AC) resulting from applying a sinusoidal potential (E_t) of small amplitude (typically 10 mV).

Suppose that we apply a sinusoidal potential excitation E_t:

$$E_t = E_0 \cdot \sin(\omega \cdot t)$$

where E_t is the potential at time t, E_0 is the amplitude of the signal, and $\omega = 2\pi f$ is the radial frequency; f is the frequency expressed in Hertz (Hz).

The response to this potential is an AC current signal with a current intensity I_t also depending on t, with the same frequency but with an amplitude I_0 and a phase angle ϕ depending on the impedance of the system (as represented in Fig. 1.33).

$$I_t = I_0 \cdot \sin(\omega \cdot t + \phi)$$

An expression analogous to Ohm's law allows us to calculate the impedance of the system, expressed in terms of a magnitude Z_0, and a phase shift ϕ

$$Z = \frac{E_t}{I_t} = \frac{E_0 \cdot \sin(\omega \cdot t)}{I_0 \cdot \sin(\omega \cdot t + \phi)} = Z_0 \cdot \frac{\sin(\omega \cdot t)}{\sin(\omega \cdot t + \phi)}$$

According to Euler's expression:

$$\exp(j\phi) = \cos\phi + j\sin\phi$$

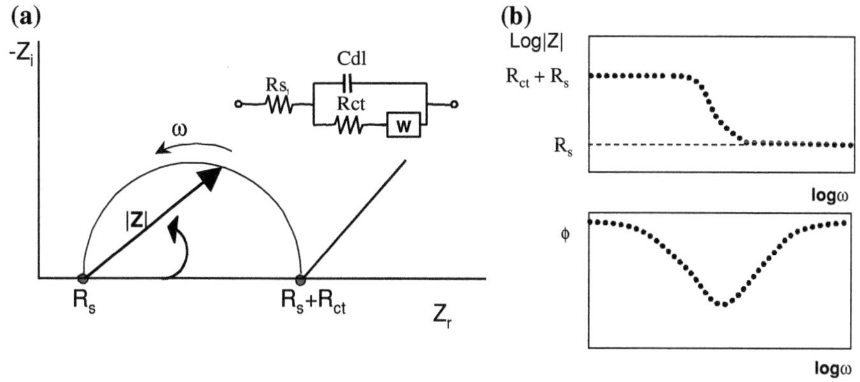

Fig. 1.34 Different representations of the impedance data. **a** Nyquist plot representing the impedance vector in the complex plane; **b** bode diagrams and the typical equivalent circuit at the electrochemical interface (Randles circuit)

A common way to represent the impedance vector model is to use complex notation:

$$E(t) = E_0 \cdot \exp(j\omega \cdot t)$$

$$I(t) = I_0 \cdot \exp(j\omega \cdot t - j\phi)$$

Impedance then is represented as:

$$Z = \frac{E}{I} = Z_0 \exp(j\phi) = Z_0 (\cos\phi + j \sin\phi) = Z_r + jZ_i$$

where Z_r is the real part of the impedance and Z_i the imaginary part.

There are many ways to plot impedance data. One of the more common impedance representations is the 'Nyquist plot' in which the imaginary part or the impedance is plotted versus the real part. It provides visual insight into the system dynamics at the electrochemical interface. As shown in Fig. 1.34, such plot exhibits a kind of semicircle profile plus a linear region. The points at which the impedance data cut the real impedance axe, represent resistance values. In the typical plot of the figure R_{ct} is the charge-transfer resistance, which is inversely proportional to the rate of electron transfer and consequently provides us information about the easiness for electron transfer at the electrode interface (a rough estimation of R_{ct} is related with the width of the semicircle); C_{dl} is the double-layer capacitance and can be obtained from the maximum value of the imaginary part of impedance at the semicircle; R_s is the electrolyte resistance and can be extracted from impedance data at the higher frequencies; W is the Warburg impedance and is identified with the linear portion of the impedance spectra that appears at the lower frequencies. The Warburg impedance arises from mass-transfer limitations between the solution and electrode surface and can be used to measure effective diffusion coefficients. It can be modelled as a frequency dependent reactance with equivalent real and imaginary components $W = \sigma \cdot (\omega)^{\frac{1}{2}} \cdot (1 - j)$ where ω = radial frequency and σ = Warburg coefficient (constant for a defined system).

Another way to represent the impedance results is by using the 'Bode diagram' where the modulus of the impedance (Log|Z|) and the phase angle (ϕ) between the AC potential and the AC current as a function of the frequency (log ω) are plotted (Fig. 1.34). In this diagram, the impedance data which are frequency independent represent the behaviour of the resistive processes (phase angles close to 0) whereas the ones that are dependent on the frequency are more related to capacitive or diffusive processes (phase angles between −90° or −45° respectively). Thus the impedance spectra can give us a broad overview of the different processes taking place at the electrochemical interface (capacitive, resistive, diffusion effects) and which one is dominating more at a specific range of frequencies.

It has long been recognized that the impedance of solid electrodes usually deviates from purely capacitive behaviour; this is empirically modelled as a constant phase element (CPE) [255, 256]. The complex impedance of a CPE is given by:

$$\frac{1}{Z_{CPE}} = (j\omega)^n C$$

where ω is the radial frequency (expressed in rad s^{-1}), C condenser capacitance and $0.5 < n < 1$ ($n = 1$ corresponds to the ideal capacitor and $n = 0.5$ corresponds to a Warburg element).

The use of CPE elements is widely extended in the experimentalists since capacitors often do not behave ideally and can be explained mathematically by dispersion in local capacitance values. Microscopic roughness can cause this effect [257], but microscopic chemical inhomogeneities and ion adsorption play an even larger role [258–260]. Solid electrodes can be expected to have a certain amount of CPE behaviour, and thus modelling the electrode-solution interface as purely capacitive is often simplistic and can reduce the quality of data fitting.

1.4.1.3 Electric Field Detection

Electronic detection based on electric field is quite promising for the bio-world since most biological processes involve electrostatic interactions and charge transfer which can be sensitively captured by FET devices, enhancing at the same time the merging of biology and electronics. Under this context, we have mentioned before that the CNT-FET configuration specifically holds promise for the development of (bio)chemical sensors since the semiconducting CNTs behave as conducting channels in the transistor approach, in direct contact with the environment and with all the current flowing at the surface of the tubes. All these features make the conductance of these devices be extremely sensitive to very small changes in the environment and to the interaction with chemical and biological species.

Many papers can be found in literature in which the capabilities of CNT-FETs have been explored. The devices have been found to be sensitive to different gases such as NH_2 or NO_2 [261] and due to size compatibility of the nanoscale devices, they have also shown to be very appropriate for single (bio)molecule detection [262].

As also pointed out before, one can get two different device architectures, one based on an individual CNT or on a random CNT network. On the former type of device, some irreproducibilities can be obtained from device to device due to the geometry-dependent variation in the electronic characteristics of individual nanotubes or in variations of the Schottky barriers. In the latter configuration, current flows along several conducting channels that determine the overall device conductance. In this architecture the device operation depends upon the density of nanotubes and also on the extent of metallic nanotubes which do not exert conductance modulation. However, the advantage of such device, though less sensitive than single nanotube devices, resides in the fact that they offer more reproducibility and manufacturability.

The presence of an immobilized biomolecule or the reaction between biomolecules (ligand-receptor) can be monitored in real-time by examining the conductance change in the device. The technique can become sensitive enough to follow conformational changes of biomolecules or to capture electronic data that may produce electronic signatures specific to a biological process. This technique can be termed as a label-free methodology with detection limits that can be down to the picomolar range. Indeed, one can consider that the CNT in the FET can act a "channel modulation label" to sense changes in their immediate environment.

Although a big amount of biosensing studies has been carried out by using carbon nanotube transistors, the physical mechanism behind the sensing process is still not so clear and the lack of a good understanding of the sensing mechanism can hamper the further exploitation of these promising nanosensors. Following, it will be shown how one can use the I–V_G curves as a tool to illustrate the changes in the CNT conductance profile from the interaction of (bio)chemical species with the CNT-FET. The evaluation of the I/V_G profiles can give insights of the different mechanisms that could take place when using CNT-FETs. In Fig. 1.35a we can see the change of the current as a function of V_G in the case of electrostatic gating by adsorbed species on the CNT. Such species can induce a partial charge transfer or doping in the CNT and thus shifting the I/V_G curve. For instance, the adsorption of positively charged species induces additional negative charge in the SWNT, thus producing n-doping in the SWNT and shifting the I/V_G curve towards more negative voltages. The opposite holds for negatively charged adsorbents.

Figure 1.35b shows that during the interaction with (bio)chemical species changes in the Schottky barrier at the metal/CNT contacts can be produced. Adsorbed species at the contacts can modulate the local work function of the contacts and thus the band alignment or bending.

Figure 1.35c shows the effect of a reduced gate efficiency which may occur when the gate capacitance is reduced due to low permittivity ε of adsorbed biomolecules relative to the electrolyte. However, this effect can become only important in case of near-full coverage of the CNT by the species and under the disputable assumption that ions cannot permeate through the adsorbed biomolecular layer.

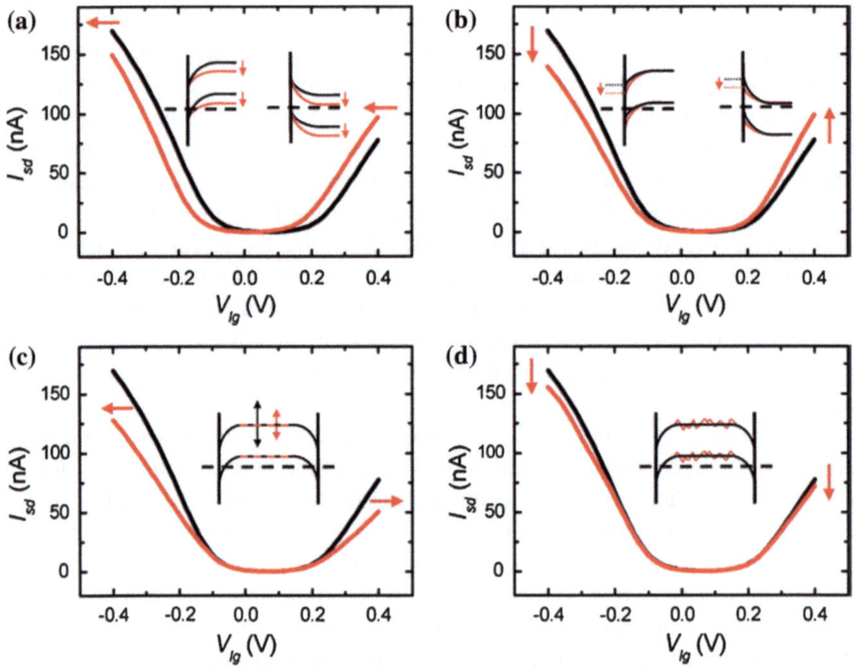

Fig. 1.35 Calculated I–V_G curves before (*black*) and after (*red*) protein adsorption for four different sensing mechanisms. The bias voltage is 10 mV. **a** Electrostatic gating effect corresponding to a shift of the semiconducting bands downward; **b** Schottky barrier effect that corresponds to a change of the difference between metal and CNT work function; **c** capacitance mechanism for 90 % coverage of CNT with protein; **d** mobility mechanism that corresponds to a mobility reduction of charge carriers to a mere 2 % of the initial value. Reprinted wit permission from Ref. [263]. Copyright (2008) American Chemical Society

Finally, Fig. 1.35d shows the effect of reduced carrier mobility by adsorption of species which produce scattering centres. That effect induces conductance suppressing in both p and n-branches [154, 263].

So far we have presented different processes that can induce changes in the characteristics of the CNT-FET device. However, the experimental identification of the sensing mechanisms according to these features is not trivial and a certain ambipolar character of the FET is important to distinguish unequivocally the sensing mechanism, a feature that sometimes is not easy to get.

Another important issue for design and applications of CNT sensors is related with the region of the device where (bio)molecule adsorption causes significant conductance changes. There have been some reports which suggest that the work function modulation is the dominant sensing mechanism. That implies that the sensitive region is limited to the nanoscale contact regions. However, there have also been some papers that also clearly indicated that in addition, strong electrostatic gating reliably occurs along the bulk of CNT channel. This latter issue has been confirmed by passivating the contacts with some insulating layer made from polymethylmethacrylate (PMMA).

Some authors claim that the most common mechanisms can be explained by a combination of electrostatic gating and Schottky barrier effects.

As mentioned in previous section, the typical gate configuration for CNT-FET biosensors is based on a bottom gate configuration which is usually operated under dry conditions. In such device, incubation and specific binding of chemical or biological entities is performed in physiological conditions. Then, the removal of weakly bound species is performed by thorough washing steps and finally, the acquisition of the sensor response is taken in dry conditions. Another configuration is the liquid-gate transistor in which the entire device is immersed and operated in buffer solution with a nearby reference electrode also immersed in the liquid which is held at a desired gate voltage. This kind of configuration allows following up in situ and in real time the (bio)chemical interactions or (bio)recognition processes in a friendly environment for the (bio)molecules in contrast to the dry conditions. The operating voltage regime for liquid-gate needs to be confined away from any electrochemical side reactions and normally the gate potential range is much narrower than in the case of the back-gate FET but the transconductance of FETs in liquids is much higher which make them more sensitive.

Another critical point in the CNT-FET (bio)sensors is the Debye length (λ_D). The Debye length is the typical distance required for screening the surplus charge by the mobile carriers present in a material (is the characteristic thickness of the double layer formed at the conducting CNT channel). The CNT-FET can only be sensitive to the detection of a (bio)molecule interaction or biorecognition event if such processes are taken place inside the Debye length. The thickness of the Debye length varies as the inverse square root of the ionic strength (I). For this reason, it results very important to adjust the ionic strength conditions. Figure 1.36 shows a hybridized DNA strand and the Debye length (as a function of the ionic strength) and how it scales with the length of biomolecule. By decreasing the ionic strength, the Debye length is bigger and the hybridization process can be detected in a major extent [264].

Thus, the issue with the Debye length restricts the detection to the size of the biomolecule or to the distance at which the biorecognition event is taken place. For instance, some authors have found that in some cases, the electronic protein detection based on an antigen/antibody protocol is not so sensitive because the immune-recognition process exceeds the limits of the Debye length. In such cases the lack of sensitivity has been overcome by using aptamers. Figure 1.37 shows an scheme showing such effect.

1.4.2 Surface Characterization: SEM, TEM, AFM

Microscopes have changed the way we approach the micro/nanoscopic world and analyse surface phenomena in physics, chemistry, biology, biochemistry and engineering. First designed as techniques to image surfaces with unprecedented spatial resolution, they have extended their capabilities to follow atomic and (bio)molecular processes in different environments and to probe local chemical and physical

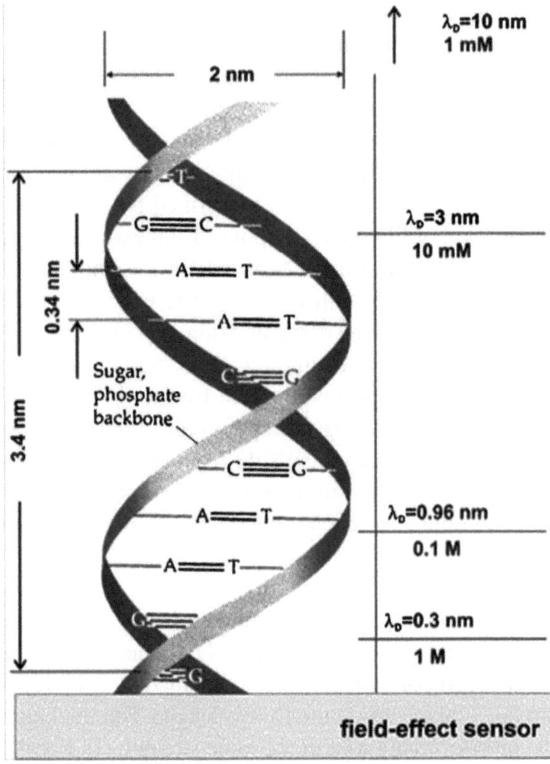

Fig. 1.36 Illustration showing the changes of the Debye length as a function of the ionic strength and in comparison with a double strand DNA

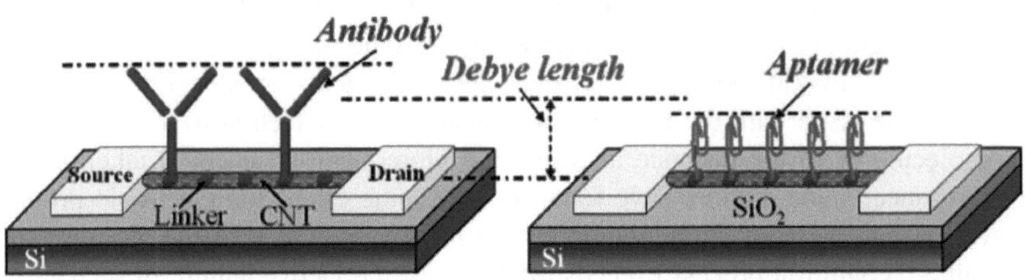

Fig. 1.37 Schematic representation of label-free protein biosensors based on CNT-FETs: **a** antibody-modified CNT-FET, **b** aptamer modified CNT-FET. It can be observed that in case (**a**) the antibody is much larger in size than the Debye length. In such case the charges of the bound protein may be screened by the double layer and their effect on the equilibrium carrier distribution would then vanishingly small. On the other hand, aptamers enable sensitive detection possibilities, partly derived from their small size. Reproduced from Ref. [264]

properties, producing a wealth of related techniques. They have also evolved from a qualitative technique to a more quantitative one [265].

In few lines, the surface characterization techniques used to follow all the studies of this thesis work such as Scanning Electron Microscopy, Transmission Electron Microscopy and Atomic Force Microscopy have been described.

1.4.2.1 Scanning Electron Microscopy

Scanning Electron Microscopy (SEM) is one of the most versatile methods for observation and characterization of heterogeneous organic and inorganic materials on a nanometre (nm) to micrometer (μm) scale. In SEM, an electron beam is moved in a raster pattern across the surface of a sample or is static to obtain an analysis at one position. The beam interacts with the sample surface, producing a number of different signals (secondary electrons, backscattered electrons, characteristic X-rays and other photons of various energies), which after amplification can be analysed to provide useful information about surface topography, composition, crystallography, etc. The imaging signals of greatest interest are the secondary and backscattered electrons (BSE) because these vary primarily as a result of differences in surface topography or provide information about the distribution of different elements in the sample (the intensity of the BSE signal is strongly related to the atomic number of the specimen). The analysis of the characteristic x-radiation emitted from samples can yield both qualitative identification and quantitative elemental information from regions of a specimen nominally 1 μm in diameter and 1 μm in depth under normal operating conditions [Energy-dispersive X-ray spectroscopy (EDS)]. Electromagnetic lenses are used to focus and deflect the electron beam. The primary motivation for using SEM instead of light microscopy is related to the fact that electrons have a much shorter wavelength than light (higher resolution) obtaining topographic images in the magnification ranges 10–10,000X and that SEM uses a larger focal length (greater depth of focus) yielding a characteristic three-dimensional appearance useful for understanding the surface structure of a sample [266, 267].

Scanning electron microscopy and sometimes combined with other techniques which help to gain resolution, is used to determine impurities such as amorphous carbon or catalyst particles that coexist with bundles of SWNTs in the sample.

For instance, Fig. 1.38 shows a SEM image of non-purified MWNTs material in the form of spaghetti grown with the CVD technique. Next to the nanotubes, the bright spots indicate the presence of catalyst particles in the sample.

1.4.2.2 Transmission Electron Microscope

In a conventional transmission electron microscope, a thin specimen is irradiated with an electron beam (acceleration voltage 100–300 keV) of uniform current density. Under high vacuum, electrons interact strongly with the specimen as it passes through by elastic and inelastic scattering. The specimen must therefore be very thin, typically of the order of 5–100 nm for 100 keV electrons, depending on the density and elemental composition of the object and the resolution desired.

Electrons are emitted in the electron gun by thermionic, Schottky, or field emission. A three or four-stage condenser lens system permits variation of the

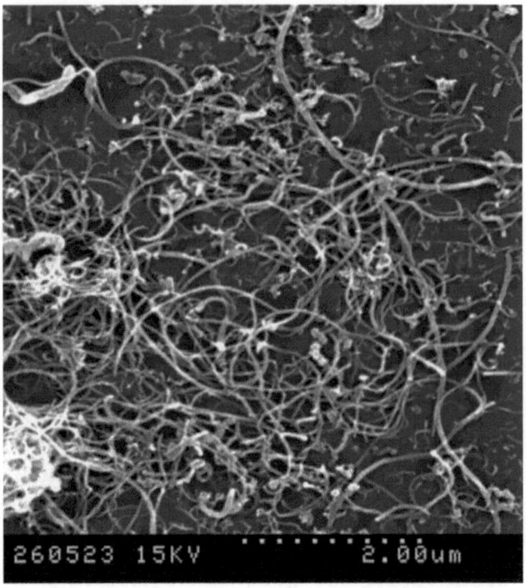

Fig. 1.38 SEM image of non- purified MWNTs (as-prepared)

illumination aperture and the area of the specimen illuminated. Because the TEM is a multiple lens system, the electron intensity distribution behind the specimen is imaged with three to eight lenses, onto a fluorescent screen. The image can be recorded by direct exposure of photographic emulsion or an image plate inside the vacuum, or digitally via fluorescent screen coupled by a fibre-optic plate to a CCD camera [268].

The aberrations of the objective lens are so great that it is necessary to work with very small objective apertures (10–25 mrad) to achieve resolution of the order of 0.1–0.3 nm. This small resolution has proved extremely valuable in the examination of biological ultra-structures such as DNA and viruses, and the structure of materials such as grain boundary properties in metallic specimens, and failures in semiconductor devices.

Crystalline materials diffract electrons due to their short wavelengths. A parallel beam of electrons passing through a regular spaced crystal lattice in the sample holder of a TEM will form a diffraction pattern in the back focal plane of the objective lens. This can be projected onto a viewing screen or recorded on film for measurement. Study of these diffraction patterns is very useful to explain the structure of materials. As well as SEM, TEMs are equipped with elemental analysis capabilities.

Transmission electron microscopy is a useful tool for our work because can determine whether the nanotubes are single-layer or multilayer, characterize lengths and diameters and identify the presence of catalyst particles, amorphous carbon or structural defects in the nanotubes grown.

As an example, in Fig. 1.39a we observe different MWNTs. In addition to nanotubes, catalyst particles can be observed as black dots in the picture. In picture b), we observe a catalyst particle covered by graphitic layers.

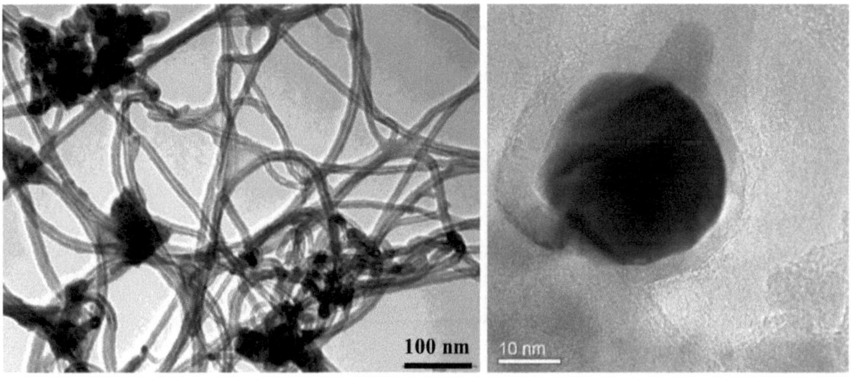

Fig. 1.39 **a** TEM image of as-prepared MWNTs; **b** catalyst particle covered by graphitic layers

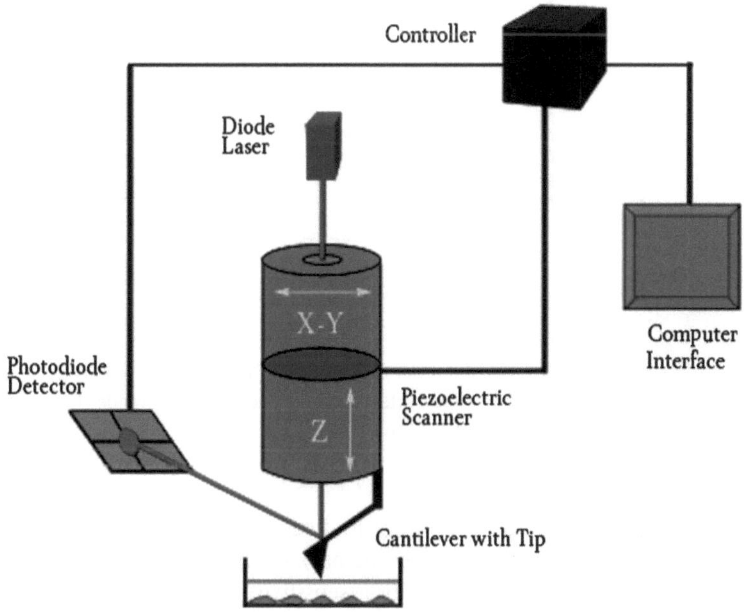

Fig. 1.40 AFM operation. The interactions between atoms in the tip and atoms of the surface can be monitored by controlling the oscillations of the cantilever through the deviation of a laser beam whose reflection impacts on a photodetector

1.4.2.3 Atomic Force Microscopy

The principle of AFM relies on the use of a tip mounted on a cantilever which is brought into close proximity to the surface where intermolecular forces acting between tip and sample cause the cantilever to bend (Fig. 1.40). The system cantilever plus tip, is connected with a xyz piezoelectric element and is moved line

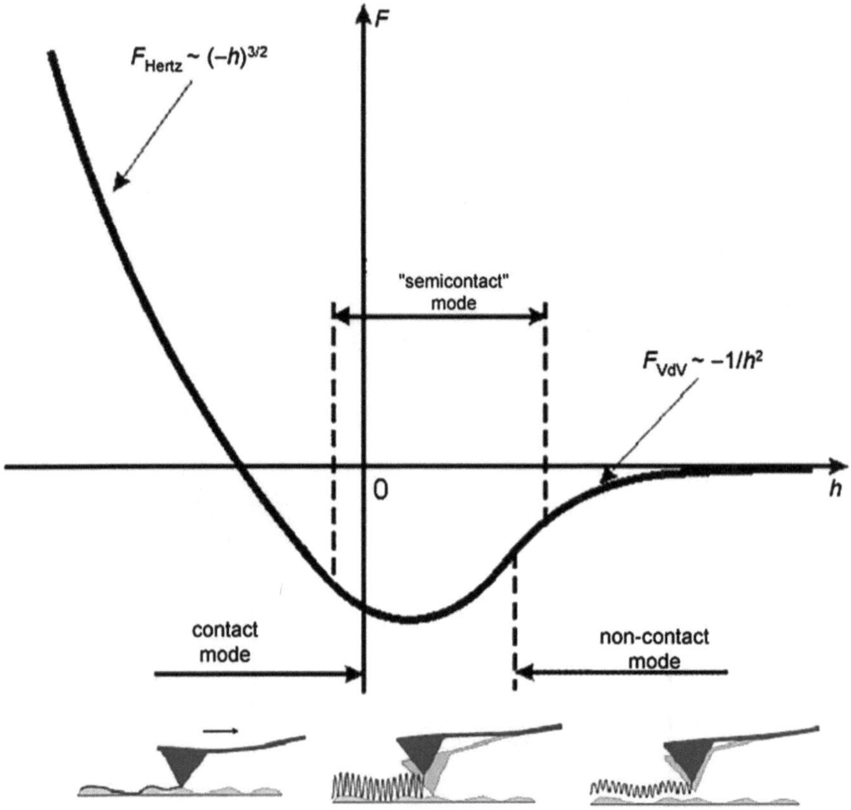

Fig. 1.41 Different modes of operation of an AFM reflected in a curve of force versus distance. Non-contact mode (attractive forces), intermittent mode or "semicontact" mode (forces of attraction/repulsion) and contact mode (repulsive forces)

by line over the sample surface. In order to detect the bending, which is as small as 0.01 nm, a laser beam is focused on the back of the cantilever. From there, the laser beam is reflected towards a position-sensitive photodetector. Depending on the cantilever deflection, the position of the reflected beam changes. The photodetector converts this change in an electrical signal which gives the topographic image of the surface. This technique can also achieve very good spatial resolution (x, y: 2–10 nm, z: 0.1 nm) and resolve interaction forces with piconewton sensitivity.

There are several types of forces that can be registered: Van der Waals forces, repulsive forces, magnetic, electrostatic forces, etc. There are also different ways to register these forces, either by non-contact between tip and sample (where mainly attractive forces are measured), by intermittent contact with the sample (where both forces attraction and repulsive are measured) or by full contact between tip and sample (where repulsive forces are measured) (Fig. 1.41).

The first development of the AFM was based on measuring the repulsive forces by contacting the sample with the tip. Since in this standard operating mode the tip is mechanically in contact with the surface, lateral forces are also present and exert a torsional movement of the cantilever as the tip scans the surface. This information can be also collected since the photodetector is a quadrant which measures both normal cantilever bending (due to topographic effects) and torsion. Thus, lateral forces, usually due to differences in friction forces on the sample, can be measured with AFM simultaneously with the topographic images.

The dragging motion of the tip in contact AFM, combined with adhesive and lateral forces, can cause substantial damage to soft samples. To alleviate this problem, dynamic AFM modes, in which the topography is monitored by the changes in the oscillation of a vibrating tip, are preferable [269] (e.g. non-contact or intermittent contact AFM mode). Very quickly, it was realized that the dynamic AFM modes could bring about additional valuable information. The existence of several parameters such as the oscillation amplitude, frequency and phase shift which are sensitive to the tip-sample interactions could be used to extract quantitative and qualitative information about material properties at nanometre scale. This sophistication was even increased by applying bias potential between tip and sample and by using AFM tips of a different nature.

The main advantages of AFM are: the capacity to work in air, vacuum and liquid medium (for biological samples), and analyse both insulating materials and conductors of electricity, in addition to allowing dynamic monitoring of physicochemical and biological processes in situ.

As stated, AFM provides topographical information; as an example, the AFM is very useful to visualize isolated nanotubes grown directly on silicon substrates by CVD method. Although the resolution is not as high as the one found with the TEM, the diameter can be estimated fairly accurately measuring the heights of the CNTs; one can also measure the lengths of the nanotubes, metal particles and to observe the distribution of disordered forms of carbon. Thus, the atomic force microscopy is an important tool in the investigation of nanotubes.

Figure 1.42 shows 2 μm × 2 μm acquired AFM topographic images where the height of an individual single-walled carbon nanotube is estimated, few nanometres, grown on silicon substrate by the CVD technique. The diameter of the particle catalyst, 3–6 nm, is also observed.

As anticipated and beyond topographic information, the AFM can access local physiochemical properties at the nanometre scale depending on the mode of operation and the nature of the tip. For example one can acquire information about the mechanical properties (adhesion, elasticity, friction), magnetic and electronic of the sample. One can also infer chemical properties. Note that the AFM can not provide chemical microanalysis determining elements or the nature of the composition of a sample as in the case of SEM and TEM but in some cases can infer chemical properties.

By using the AFM in the current sensing mode, conductance mappings of surfaces are also performed. This technique consists in the use of a conductive tip operating in standard contact mode. As a voltage is applied between tip and

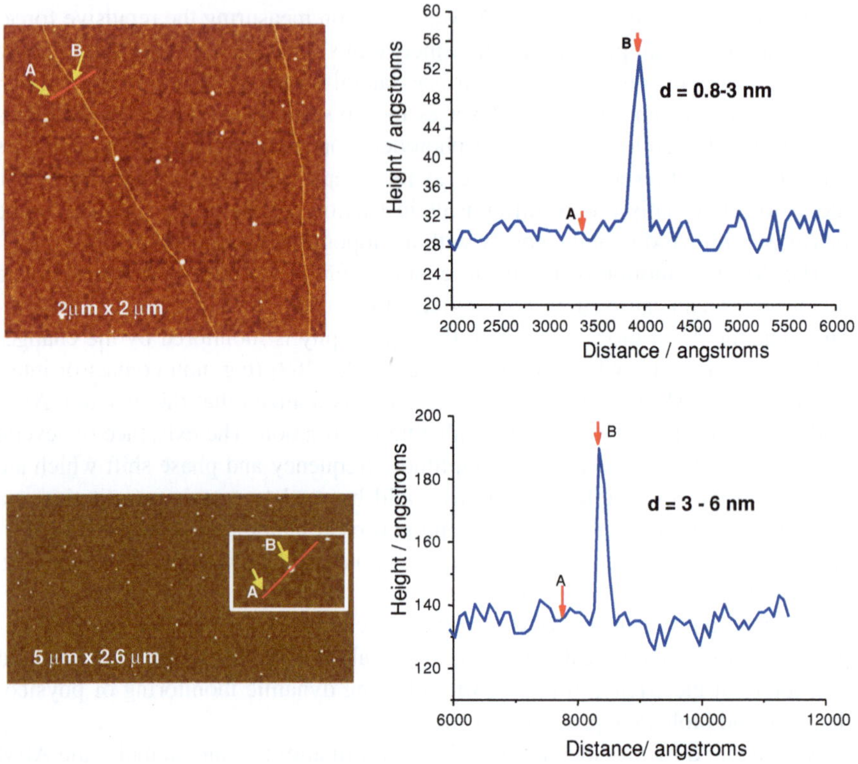

Fig. 1.42 AFM images and topographic profiles where the diameter of a nanotube and a catalyst particle are estimated

sample, a current is generated, the intensity of which will depend on the sample nature. Thus, a map of current distribution along the sample is measured simultaneously with the surface topography (Fig. 1.43).

Current sensor mode is suitable for samples with low conductance or composite materials containing insulating regions. This sensor mode has been an interesting tool for characterizing the different phases and their distributions in the composite fabrication.

1.4.3 Other Characterization Techniques: Raman, XPS

Other characterization techniques have been used in order to have more insights on the structural characteristics of the nanotubes, electrodes, or to verify functionalization strategies on carbon nanotubes. These are Raman spectroscopy and X-ray photoelectron spectroscopy (XPS).

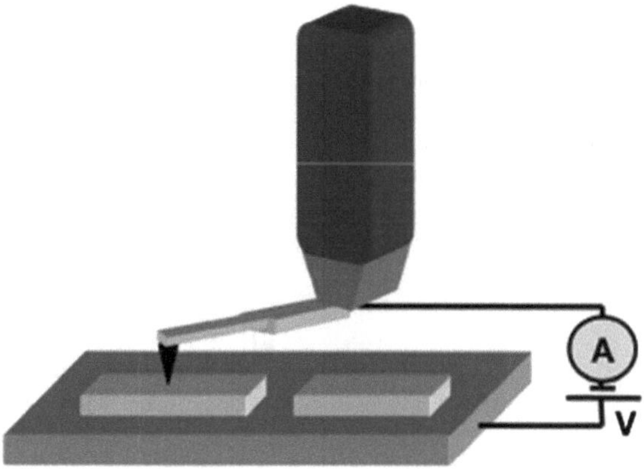

Fig. 1.43 Current sensor mode scheme

1.4.3.1 Raman Spectroscopy

Raman spectroscopy is a spectroscopic technique used to study vibrational, rotational, and other low-frequency modes in a system [270]. It relies on inelastic scattering of monochromatic light, usually from a laser in the visible, near infrared, or near ultraviolet range.

The laser light interacts with molecular vibrations or phonons which make the energy of the laser photons shift up or down. The shift in energy gives information about the vibrational modes in the system and becomes a complementary tool of infrared spectroscopy.

The Raman effect occurs by interaction of radiation with the electron cloud and the bonds of molecules or crystal. It can be explained as follows, when radiation impinges on the sample, the molecule is photoexcited from the ground state to a virtual energy state. When the molecule relaxes it emits a photon and it returns to a different rotational or vibrational state. The difference in energy between the original state and this new state leads to a shift in the emitted photon's frequency away from the excitation wavelength. If the final vibrational state of the molecule is more energetic than the initial state, then the emitted photon will be shifted to a lower frequency so that the total energy of the system remains balanced. This shift in frequency is designated as a Stokes shift. If the final vibrational state is less energetic than the initial state, then the emitted photon will be shifted to a higher frequency, and this is designated as an Anti-Stokes shift. Raman scattering is an example of inelastic scattering because of the energy transfer between the photons and the molecules during their interaction.

To exhibit Raman effect in molecules, a change in the molecular polarization potential—or amount of deformation of the electron cloud—with respect to the vibrational coordinate is required. The amount of the polarizability change will determine the Raman scattering intensity. The pattern of shifted frequencies is determined by the rotational and vibrational states of the sample.

Ramon is a very useful technique to characterize CNTs. For instance, Raman spectra present different features being all sensitive to (n, m), such as the radial

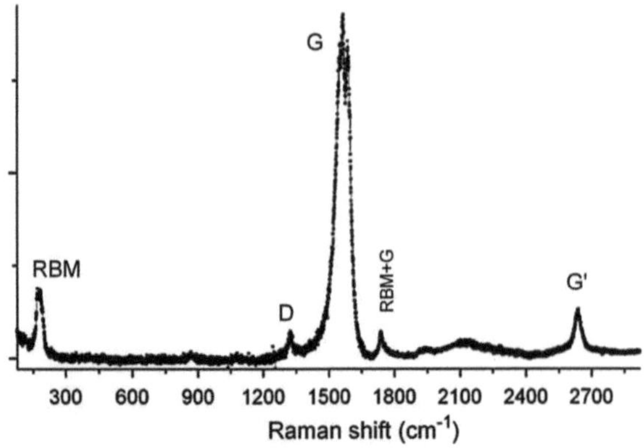

Fig. 1.44 Raman spectrum of well-dispersed single-wall carbon nanotubes (Hipco): *RBM* radial breathing mode, *D* D mode, *G* G mode, *RMB* + *G* combination of RBM and G modes, G' double-phonon scattering of the D mode

breathing mode (RBM), where all the carbon atoms are moving in-phase in the radial direction, the G-band where neighbouring atoms are moving in opposite directions along the surface of the tube as in 2D graphite, the dispersive disorder induced D-band and its second-order related harmonic G′-band. The so-called G-band, around 1,580cm^{-1}, is a characteristic feature of the graphitic layers. The second characteristic mode is a typical sign for defective graphitic structures (D-band, around 1,350 cm^{-1}). The comparison of the ratios of these two peaks intensities gives a measure of the quality of the bulk samples. If these both bands have similar intensity this indicates a high quantity of structural defects. Therefore, I_D/I_G ratio was extracted as a parameter to determine the structural quality of the carbon materials.

One important feature in the Raman spectrum of SWNT is the Radial Breathing Mode (RBM), which is usually located between 75 and 300 cm^{-1} from the exciting line; an illustration of the spectrum resulting from this mode is displayed in the Fig. 1.44. The frequency of the RBM is very sensitive to the diameter of the nanotubes being directly linked to the reciprocal of the nanotube diameter (d_t) and is expressed as $\varpi_{RBM} = C/d_t$ (cm^{-1}) (C = 248 cm^{-1} for isolated SWNTs on a SiO_2 substrate [271]). However, non isolated SWNTs are subject to inter-tube interactions which increase the frequency of the RBM [65, 272–274].

1.4.3.2 Spectroscopy X-ray Photoelectron

Spectroscopy X-ray photoelectron (XPS) is a very sensitive and non-destructive surface technique that provide quantitative information on the chemical status of a system. The sample is irradiated with a monochromatic beam of X-rays (XPS), the photon is absorbed by an atom in a molecule or solid, leading to ionization and the emission of a core (inner shell) electron and the emission energy of the photoelectrons is analysed.

For each and every element, there will be a characteristic binding energy associated with each core atomic orbital i.e. each element will give rise to a characteristic set of peaks in the photoelectron spectrum at kinetic energies determined by the photon energy and the respective binding energies.

The presence of peaks at particular energies therefore indicates the presence of a specific element in the sample under study—furthermore, the intensity of the peaks is related to the concentration of the element within the sampled region. Thus, the technique provides a quantitative analysis of the surface composition and is sometimes known by the alternative acronym, Electron Spectroscopy for Chemical Analysis (ESCA).

With this technique, the empirical formula, chemical state and electronic state of the elements that exist within a material can also be measured.

References

1. Weiss J, Takhistov P, McClements DJ (2006) J Food Sci 71:R107–R116
2. Oberlin A, Endo M, Koyama T (1976) J Cryst Growth 32:335–349
3. Kroto HW, Heath JR, O'Brien SC, Curl RF, Smalley RE (1985) Nature 318:162–163
4. Iijima S (1991) Nature 354:56–58
5. Endo M, Strano M, Ajayan P (2008) In: Jorio A, Dresselhaus G, Dresselhaus MS (eds) Carbon nanotubes, vol 111. Springer, Berlin, pp 13–61
6. Saito R, Dresselhaus G, Dresselhaus MS (eds) (1998) Physical properties of carbon nanotubes. Imperial College Press, London
7. Louie S (2001) In: Dresselhaus M, Dresselhaus G, Avouris P (eds) Carbon nanotubes, vol 80. Springer, Berlin, pp 113–145
8. Ouyang M, Huang J-L, Lieber CM (2002) Acc Chem Res 35:1018–1025
9. Avouris P (2002) Acc Chem Res 35:1026–1034
10. Balasubramanian K, Burghard M (2005) Small 1:180–192
11. Wildöer JWG, Venema LC, Rinzler AG, Smalley RE, Dekker C (1998) Nature 391:59–62
12. Odom TW, Huang JL, Kim P, Lieber CM (1998) Nature 391:62–64
13. Krstić V, Roth S, Burghard M (2000) Phys Rev B 62:R16353
14. Kong J, Yenilmez E, Tombler TW, Kim W, Dai H, Laughlin RB, Liu L, Jayanthi CS, Wu SY (2001) Phys Rev Lett 87:106801
15. Heinze S, Tersoff J, Martel R, Derycke V, Appenzeller J, Avouris P (2002) Phys Rev Lett 89:106801
16. Hamon MA, Itkis ME, Niyogi S, Alvaraez T, Kuper C, Menon M, Haddon RC (2001) J Am Chem Soc 123:11292–11293
17. Han J (2005) In: Meyyappan M (ed) Carbon nanotubes: science and applications. CRC Press, London, p 1
18. Srivastava D (2005) In: Meyyappan M (ed) Carbon nanotubes: science and applications. CRC Press, London, p 25
19. Stone AJ, Wales DJ (1986) Chem Phys Lett 128:501–503
20. Roh S, Lee J, Jang M (2010) J Nanomater 2010:6
21. Liew KM, He XQ, Wong CH (2004) Acta Mater 52:2521–2527
22. Duplock EJ, Scheffler M, Lindan PJD (2004) Phys Rev Lett 92:225502
23. Picozzi S, Santucci S, Lozzi L, Valentini L, Delley B (2004) J Chem Phys 120:7147–7152
24. Dresselhaus MS, Dresselhaus G, Avouris P (eds) (2001) Carbon nanotubes: synthesis, structure, properties, and applications, vol 80. Springer, Berlin
25. Suenaga K, Wakabayashi H, Koshino M, Sato Y, Urita K, Iijima S (2007) Nat Nano 2:358–360
26. Salvetat J-P, Briggs GAD, Bonard J-M, Bacsa RR, Kulik AJ, Stockli T, Burnham NA, Forro LL (1999) Phys Rev Lett 82:944
27. Cornwell CF, Wille LT (1997) Solid State Commun 101:555–558
28. Ruoff RS, Lorents DC (1995) Carbon 33:925–930

29. Yu M-F, Files BS, Arepalli S, Ruoff RS (2000) Phys Rev Lett 84:5552
30. Yu M-F, Lourie O, Dyer MJ, Moloni K, Kelly TF, Ruoff RS (2000) Science 287:637–640
31. Demczyk BG, Wang YM, Cumings J, Hetman M, Han W, Zettl A, Ritchie RO (2002) Mater Sci Eng, A 334:173–178
32. Shen W, Jiang B, Han BS, Xie S–S (2000) Phys Rev Lett 84:3634
33. Mitchell LA, Gao J, Wal RV, Gigliotti A, Burchiel SW, McDonald JD (2007) Toxicol Sci 100:203–214
34. Niyogi S, Hamon MA, Hu H, Zhao B, Bhowmik P, Sen R, Itkis ME, Haddon RC (2002) Acc Chem Res 35:1105–1113
35. Chen Z, Thiel W, Hirsch A (2003) ChemPhysChem 4:93–97
36. Burghard M (2005) Surf Sci Rep 58:1–109
37. McCreery RL (2008) Chem Rev 108:2646–2687
38. McCreery RL (1991) In: Bard AJ (ed) Electroanalytical chemistry, vol 17. Dekker, New York, 1991
39. Merkoçi A (2006) Microchim Acta 152:157–174
40. Wang J (2005) Electroanalysis 17:7–14.
41. Pumera M, Sánchez S, Ichinose I, Tang J (2007) Sens Actuators B Chem 123:1195–1205
42. Agüí L, Yáñez-Sedeño P, Pingarrón JM (2008) Anal Chim Acta 622:11–47
43. Heller I, Kong J, Williams KA, Dekker C, Lemay SG (2006) J Am Chem Soc 128: 7353–7359
44. Gooding JJ, Wibowo R, Liu JQ, Yang W, Losic D, Orbons S, Mearns FJ, Shapter JG, Hibbert DB (2003) J Am Chem Soc 125:9006–9007
45. Nugent JM, Santhanam KSV, Rubio A, Ajayan PM (2001) Nano Lett 1:87–91
46. Moore RR, Banks CE, Compton RG (2004) Anal Chem 76:2677–2682
47. Banks CE, Davies TJ, Wildgoose GG, Compton RG (2005) Chem Commun 7:829–841
48. Banks CE, Compton RG (2006) Analyst 131:15–21
49. Wildgoose GG, Banks CE, Leventis HC, Compton RG (2006) Microchim Acta 152:187–214
50. Banks CE, Crossley A, Salter C, Wilkins SJ, Compton RG (2006) Angew Chem Int Ed 45:2533–2537
51. Gooding JJ (2005) Electrochim Acta 50:3049–3060
52. Gooding JJ, Lai LMH, Goon IY (2009) In: Alkire RC et al (eds) Chemically modified electrodes. Wiley-VCH, Weinheim, pp 1–56
53. Yu X, Chattopadhyay D, Galeska I, Papadimitrakopoulos F, Rusling JF (2003) Electrochem Commun 5:408–411
54. Banks CE, Ji X, Crossley A, Compton RG (2006) Electroanalysis 18:2137–2140
55. Chou A, Böcking T, Singh NK, Gooding JJ (2005) Chem Commun 7:842–844
56. Misewich JA, Martel R, Avouris P, Tsang JC, Heinze S, Tersoff J (2003) Science 300:783–786
57. Chen J, Perebeinos V, Freitag M, Tsang J, Fu Q, Liu J, Avouris P (2005) Science 310:1171–1174
58. Freitag M, Martin Y, Misewich JA, Martel R, Avouris P (2003) Nano Lett 3:1067–1071
59. Itkis ME, Borondics F, Yu A, Haddon RC (2006) Science 312:413–416
60. Star A, Lu Y, Bradley K, Grüner G (2004) Nano Lett 4:1587–1591
61. Miyauchi Y, Oba M, Maruyama S (2006) Phys Rev B 74:205440
62. Iakoubovskii K, Minami N, Kim Y, Miyashita K, Kazaoui S, Nalini B (2006) Appl Phys Lett 89:173108-3
63. Dresselhaus MS, Dresselhaus G, Jorio A (2007) J Phys Chem C 111:17887–17893
64. Heller DA, Barone PW, Swanson JP, Mayrhofer RM, Strano MS (2004) J Phys Chem B 108:6905–6909
65. Oron-Carl M, Hennrich F, Kappes MM, Löhneysen HV, Krupke R (2005) Nano Lett 5: 1761–1767
66. Kim UJ, Furtado CA, Liu X, Chen G, Eklund PC (2005) J Am Chem Soc 127:15437–15445

67. Joselevich E, Dai H, Liu J, Hata K, Windle AH (2008) In: Jorio A, Dresselhaus G, Dresselhaus MS (eds) Carbon nanotubes, vol 111. Springer, Berlin, pp 101–164
68. Shi Z, Lian Y, Liao FH, Zhou X, Gu Z, Zhang Y, Iijima S, Li H, Yue KT, Zhang S-L (2000) J Phys Chem Solids 61:1031–1036
69. Moravsky AP, Wexler EM, Loutfy RO (2005) In: Meyyappan M (ed) Carbon nanotubes: science and applications. CRC Press, London, p 65
70. Meyyappan M (2005) In: Meyyappan M (ed) Carbon nanotubes: science and applications. CRC Press, London, p 99
71. Dai H (2002) Surf Sci 500:218–241
72. Dai H, Rinzler AG, Nikolaev P, Thess A, Colbert DT, Smalley RE (1996) Chem Phys Lett 260:471–475
73. Veronese GP, Rizzoli R, Angelucci R, Cuffiani M, Malferrari L, Montanari A, Odorici F (2007) Phys E Low-Dim Syst Nanostruct 37:21–25
74. Kondo D, Sato S, Awano Y (2006) Chem Phys Lett 422:481–487
75. Song IK, Cho YS, Choi GS, Park JB, Kim DJ (2004) Diamond Relat Mater 13:1210–1213
76. Bower C, Zhou O, Zhu W, Werder DJ, Jin S (2000) Appl Phys Lett 77:2767–2769
77. Meyyappan M et al (2003) Plasma Sources Sci Technol 12:205
78. Ebbesen TW, Ajayan PM, Hiura H, Tanigaki K (1994) Nature 367:519
79. Chattopadhyay D, Galeska I, Papadimitrakopoulos F (2002) Carbon 40:985–988
80. Rinzler AG, Liu J, Dai H, Nikolaev P, Huffman CB, Rodríguez-Macías FJ, Boul PJ, Lu AH, Heymann D, Colbert DT, Lee RS, Fischer JE, Rao AM, Eklund PC, Smalley RE (1998) Appl Phys A Mater Sci Process 67:29–37
81. Zhou O, Shimoda H, Gao B, Oh S, Fleming L, Yue G (2002) Acc Chem Res 35:1045–1053
82. Andrews R, Jacques D, Qian D, Rantell T (2002) Acc Chem Res 35:1008–1017
83. Dillon AC, Gennett T, Jones KM, Alleman JL, Parilla PA, Heben MJ (1999) Adv Mater 11:1354–1358
84. Jurkschat K, Ji X, Crossley A, Compton RG, Banks CE (2007) Analyst 132:21–23
85. Pumera M (2007) Langmuir 23:6453–6458
86. Jones CP, Jurkschat K, Crossley A, Compton RG, Riehl BL, Banks CE (2007) Langmuir 23:9501–9504
87. Swain GM (2006) In: Zoski CG (ed) Handbook of electrochemistry. Elsevier, London, pp 431–469
88. Millan KM, Spurmanis AJ, Mikkelsen SR (1992) Electroanalysis 4:929–932
89. Kang J, Li X, Wu G, Wang Z, Lu X (2007) Anal Biochem 364:165–170
90. Tian Y, Mao L, Okajima T, Ohsaka T (2005) Biosens Bioelectron 21:557–564
91. Wang J, Cai X, Rivas G, Shiraishi H, Farias PAM, Dontha N (1996) Anal Chem 68:2629–2634
92. Millan KM, Saraullo A, Mikkelsen SR (1994) Anal Chem 66:2943–2948
93. Atanasov P, Kaisheva A, Iliev I, Razumas V, Kulys J (1992) Biosens Bioelectron 7:361–365
94. Liu S, Ye J, He P, Fang Y (1996) Anal Chim Acta 335:239–243
95. Karadeniz H, Erdem A, Caliskan A, Pereira CM, Pereira EM, Ribeiro JA (2007) Electrochem Commun 9:2167–2173
96. Pividori MI, Alegret S (2003) Anal Lett 36:1669–1695
97. Santandreu M, Céspedes F, Alegret S, Martínez-Fàbregas E (1997) Anal Chem 69:2080–2085
98. Pereira AC, Aguiar MR, Kisner A, Macedo DV, Kubota LT (2007) Sens Actuators B Chem 124:269–276
99. Wang SG, Zhang Q, Wang R, Yoon SF (2003) Biochem Biophys Res Commun 311:572–576
100. Nuzzo RG, Allara DL (1983) J Am Chem Soc 105:4481–4483
101. Levicky R, Herne TM, Tarlov MJ, Satija SK (1998) J Am Chem Soc 120:9787–9792
102. Paleček E (1996) Electroanalysis 8:7–14
103. Rampi MA, Schueller OJA, Whitesides GM (1998) Appl Phys Lett 72:1781–1783
104. Bryant MA, Pemberton JE (1991) J Am Chem Soc 113:8284–8293

105. Tammeveski K, Kikas T, Tenno T, Niinistö L (1998) Sens Actuators B Chem 47:21–29
106. Burke LD, Lyons MFG (1986) In: White RE, Bockris JOM, Conway BE (eds) Modern aspects of electrochemistry, vol 18. Plenum, New York, pp 169–248
107. O'Sullivan EJM, Calvo EJ (1987) In: Compton RG (ed) Comprehensive chemical kinetics, vol 27. Elsevier, Amsterdam, Chap 3
108. Albery WJ, Bartlett PN (1984) J Chem Soc Chem Commun 4:234–236
109. Brett CMA, Brett AMO (eds) (1993) Electrochemistry: principles, methods, and applications. Oxford Press, Oxford
110. von Sturm F (1988) Angew Chem 100:1260–1261
111. Ruschan GR, Newnham RE, Runt J, Smith E (1989) Sens Actuators 20:269
112. Esplandiu MJ, Baeza M, Olive-Monllau R, Cespedes F, Bartroli J (2011) In: Attaf B (ed) Advances in composite materials for medicine and nanotechnology. InTech, Croatia
113. Godovski DY, Koltypin EA, Volkov AV, Moskvina MA (1993) Analyst 118:997–999
114. Céspedes F, Martinez-Fàbregas E, Alegret S (1996) Trends Anal Chem 15:296–304
115. Céspedes F, Martínez-Fàbregas E, Alegret S (1993) Anal Chim Acta 284:21–26
116. Céspedes F, Martínez-Fàbregas E, Bartrolí J, Alegret S (1993) Anal Chim Acta 273: 409–417
117. Pividori MI, Merkoçi A, Alegret S (2003) Biosens Bioelectron 19:473–484
118. Wang J, Musameh M (2003) Anal Chem 75:2075–2079
119. Zhang M, Smith A, Gorski W (2004) Anal Chem 76:5045–5050
120. Mendoza Gómez E, Orozco J, Jiménez-Jorquera C, González-Guerrero AB, Calle Martín A, Lechuga LM, Fernández Sánchez C (2008) Nanotechnology 19:75102
121. Sánchez S, Pumera M, Fàbregas E (2007) Biosens Bioelectron 23:332–340
122. Pacios M, del Valle M, Bartroli J, Esplandiu MJ (2008) J Electroanal Chem 619–620:117–124
123. Pumera M, Merkoçi A, Alegret S (2006) Sens Actuators B Chem 113:617–622
124. Gong K, Yan Y, Zhang M, Su L, Xiong S, Mao L (2005) Anal Sci 21:1383–1393
125. Gavalas VG, Law SA, Ball JC, Andrews R, Bachas LG (2004) Anal Biochem 329:247–252
126. Zhang B, Xu Y, Zheng Y, Dai L, Zhang M, Yang J, Chen Y, Chen X, Zhou J (2011) Nanoscale Res Lett 6:431
127. Sainz R, Benito AM, Martínez MT, Galindo JF, Sotres J, Baró AM, Corraze B, Chauvet O, Maser WK (2005) Adv Mater 17:278–281
128. Huang J, Li X, Xu J, Li H (2003) Carbon 41:2731–2736
129. Guo M, Chen J, Li J, Tao B, Yao S (2005) Anal Chim Acta 532:71–77
130. Terzic S, Tripkovic D, Jovanovic VM, Tripkovic A, Kowal A (2007) J Serb Chem Soc 72:165–181
131. Jenkins GM, Kawamura K (1971) Nature 231:175–176
132. Milchev A, Zapryanova T (2006) Electrochim Acta 51:4916–4921
133. Merkoçi A, Pumera M, Llopis X, Pérez B, del Valle M, Alegret S (2005) Trends Anal Chem 24:826–838
134. Wang J, Li M, Shi Z, Li N, Gu Z (2002) Anal Chem 74:1993–1997
135. Lawrence NS, Deo RP, Wang J (2005) Electroanalysis 17:65–72
136. Heng LY, Chou A, Yu J, Chen Y, Gooding JJ (2005) Electrochem Commun 7:1457–1462
137. Li J, Cassell A, Delzeit L, Han J, Meyyappan M (2002) J Phys Chem B 106:9299–9305
138. Koehne J, Li J, Cassell AM, Chen H, Ye Q, Ng HT, Han J, Meyyappan M (2004) J Mater Chem 14:676–684
139. Li J, Koehne JE, Cassell AM, Chen H, Ng HT, Ye Q, Fan W, Han J, Meyyappan M (2005) Electroanalysis 17:15–27
140. Lin Y, Lu F, Tu Y, Ren Z (2003) Nano Lett 4:191–195
141. Martin-Fernandez I, Gabriel G, Rius G, Villa R, Perez-Murano F, Lora-Tamayo E, Godignon P (2009) Microelectron Eng 86:806–808
142. Zoski CG (2002) Electroanalysis 14:1041–1051
143. Heinze J (1993) Angew Chem Int Ed Engl 32:1268–1288
144. Arrigan DWM (2004) Analyst 129:1157–1165

145. Fiaccabrino GC, Koudelka-Hep M, Jeanneret S, van den Berg A, de Rooij NF (1994) Sens Actuators B Chem 19:675–677
146. Orozco J, Fernández-Sánchez C, Jiménez-Jorquera C (2010) Sensors 10:475–490
147. Morf WE, de Rooij NF (1997) Sens Actuators B Chem 44:538–541
148. Belmont C, Tercier ML, Buffle J, Fiaccabrino GC, Koudelka-Hep M (1996) Anal Chim Acta 329:203–214
149. Baker WS, Crooks RM (1998) J Phys Chem B 102:10041–10046
150. Ugo P, Moretto LM, Vezzà F (2002) ChemPhysChem 3:917–925
151. Artigas J, Jimenez C, Lemos SG, Nogueira ARA, Torre-Neto A, Alonso J (2003) Sens Actuators B Chem 88:337–344
152. Galán-Vidal CA, Muñoz J, Domínguez C, Alegret S (1997) Sens Actuators B Chem 45:55–62
153. Jessica K et al (2003) Nanotechnology 14:1239
154. Gruner G (2006) Anal Bioanal Chem 384:322–335
155. Tans SJ, Verschueren ARM, Dekker C (1998) Nature 393:49–52
156. Martel R, Schmidt T, Shea HR, Hertel T, Avouris P (1998) Appl Phys Lett 73:2447–2449
157. Derycke V, Martel R, Appenzeller J, Avouris P (2002) Appl Phys Lett 80:2773–2775
158. Cid C (2009) Doctoral thesis. Universitat Rovira i Virgili, Tarragona
159. Baughman RH, Cui C, Zakhidov AA, Iqbal Z, Barisci JN, Spinks GM, Wallace GG, Mazzoldi A, De Rossi D, Rinzler AG, Jaschinski O, Roth S, Kertesz M (1999) Science 284:1340–1344
160. Yun YH, Shanov V, Schulz MJ, Narasimhadevara S, Subramaniam S, Hurd D, Boerio FJ (2005) Smart Mater Struct 14:1526–1532
161. Yun Y, Shanov V, Tu Y, Schulz MJ, Yarmolenko S, Neralla S, Sankar J, Subramaniam S (2006) Nano Lett 6:689–693
162. Hirsch A (2002) Angew Chem Int Ed 41:1853–1859
163. Han W, Fan S, Li Q, Hu Y (1997) Science 277:1287–1289
164. Sloan J, Hammer J, Zwiefka-Sibley M, Green MLH (1998) Chem Commun 3:347–348
165. Dujardin E, Ebbesen TW, Krishnan A, Treacy MMJ (1998) Adv Mater 10:1472–1475
166. Matsui K, Pradhan BK, Kyotani T, Tomita A (2001) J Phys Chem B 105:5682–5688
167. Govindaraj A, Satishkumar BC, Nath M, Rao CNR (1999) Chem Mater 12:202–205
168. Wilson M, Madden PA (2001) J Am Chem Soc 123:2101–2102
169. Smith BW, Monthioux M, Luzzi DE (1999) Chem Phys Lett 315:31–36
170. Tasis D, Tagmatarchis N, Bianco A, Prato M (2006) Chem Rev 106:1105–1136
171. Yang WR, Thordarson P, Gooding JJ, Ringer SP, Braet F (2007) Nanotechnology 18:412001
172. Katz E, Willner I (2004) ChemPhysChem 5:1084–1104
173. Kim SN, Rusling JF, Papadimitrakopoulos F (2007) Adv Mater 19:3214–3228
174. Yin Y, Lü Y, Wu P, Cai C (2005) Sensors 5:220–234
175. Li Y, Lin X, Jiang C (2006) Electroanalysis 18:2085–2091
176. Wang L, Wang J, Zhou F (2004) Electroanalysis 16:627–632
177. Zhao L, Liu H, Hu N (2006) J Colloid Interface Sci 296:204–211
178. Du P, Liu S, Wu P, Cai C (2007) Electrochim Acta 52:6534–6547
179. Zhang L, Zhao G-C, Wei X-W, Yang Z-S (2005) Electroanalysis 17:630–634
180. Zhang L, Zhao G-C, Wei X-W, Yang Z-S (2004) Chem Lett 33:86–87
181. Zhao G-C, Zhang L, Wei X-W, Yang Z-S (2003) Electrochem Commun 5:825–829
182. Zhao G-C, Yin Z–Z, Zhang L, Wei X-W (2005) Electrochem Commun 7:256–260
183. Yin Y, Wu P, Lü Y, Du P, Shi Y, Cai C (2007) J Solid State Electrochem 11:390–397
184. Nelson DL, Lenin MMC (eds) (2000) Principles in biochemistry. Worth Publishers, New York, p 175
185. Roat-Malone RM (ed) (2002) Bioinorganic chemistry. Wiley, New York, pp 158–185
186. Fita I, Rossmann MG (1985) J Mol Biol 185:21–37
187. Murphy L (2006) Curr Opin Chem Biol 10:177–184
188. Armstrong FA, Wilson GS (2000) Electrochim Acta 45:2623–2645

189. Murthy MRN, Reid TJ, Sicignano A, Tanaka N, Rossmann MG (1981) J Mol Biol 152:465–499
190. Jouve H-M, Gouet P, Boudjada N, Buisson G, Kahn R, Duee E (1991) J Mol Biol 221:1075–1077
191. Carot ML, Torresi RM, Garcia CD, Esplandiu MJ, Giacomelli CE (2010) J Phys Chem C 114:4459–4465
192. He P, Xu Y, Fang Y (2006) Microchim Acta 152:175–186
193. Drummond TG, Hill MG, Barton JK (2003) Nat Biotech 21:1192–1199
194. Wang J, Li M, Shi Z, Li N, Gu Z (2004) Electroanalysis 16:140–144
195. Wang J, Kawde A-N, Musameh M (2003) Analyst 128:912–916
196. Fan C, Plaxco KW, Heeger AJ (2003) Proc Nat Acad Sci U S A 100:9134–9137
197. He P, Dai L (2004) Chem Commun 10:348–349
198. Bonanni A, del Valle M (2010) Anal Chim Acta 678:7–17
199. Bonanni A, Esplandiu MJ, del Valle M (2008) Electrochim Acta 53:4022–4029
200. Wohlstadter JN, Wilbur JL, Sigal GB, Biebuyck HA, Billadeau MA, Dong L, Fischer AB, Gudibande SR, Jameison SH, Kenten JH, Leginus J, Leland JK, Massey RJ, Wohlstadter SJ (2003) Adv Mater 15:1184–1187
201. Sassolas A, Blum LJ, Leca-Bouvier BD (2009) Electroanalysis 21:1237–1250
202. Jayasena SD (1999) Clin Chem 45:1628–1650
203. Ellington AD, Szostak JW (1990) Nature 346:818–822
204. Doudna JA, Cech TR, Sullenger BA (1995) Proc Nat Acad Sci U S A 92:2355–2359
205. Lee S-W, Sullenger BA (1997) Nat Biotechnol 15:41–45
206. Rusconi CP, Scardino E, Layzer J, Pitoc GA, Ortel TL, Monroe D, Sullenger BA (2002) Nature 419:90–94
207. Rusconi CP, Yeh A, Lyerly HK, Lawson JH, Sullenger BA (2000) J Thromb Haemost 84:841–848
208. Stoltenburg R, Reinemann C, Strehlitz B (2007) Biomol Eng 24:381–403
209. Hamula CLA, Guthrie JW, Zhang HQ, Li XF, Le XC (2006) Trends Anal Chem 25:681–691
210. Nimjee SM, Rusconi CP, Sullenger BA (2005) Annu Rev Med 56:555–583
211. O'Sullivan C (2002) Anal Bioanal Chem 372:44–48
212. Wang H, Yang R, Yang L, Tan W (2009) ACS Nano 3:2451–2460
213. Wang Z, Lu Y (2009) J Mater Chem 19:1788–1798
214. Di Cera E (2008) Mol Aspects Med 29:203–254
215. Fenton JW, Landis BH, Walz DA, Finlayson JS (1977) In: Lundblad RL, Fenton JW, Mann KG (eds) Chemistry and biology of thrombin. Ann Arbor Science Publishers, Michigan, pp 43–70
216. Lombardi A, De Simone G, Galdiero S, Staiano N, Nastri F, Pavone V (1999) Pept Sci 51:19–39
217. Petrera NS, Stafford AR, Leslie BA, Kretz CA, Fredenburgh JC, Weitz JI (2009) J Biol Chem 284:25620–25629
218. Tegos TJ, Kalodiki E, Daskalopoulou S–S, Nicolaides AN (2000) Angiology 51:793–808
219. Chiu T-C, Huang C–C (2009) Sensors 9:10356–10388
220. Rodríguez MC, Rivas GA (2009) Talanta 78:212–216
221. Huang J, Wu L, Yalda D, Adkins Y, Kelleher SL, Crane M, Lonnerdal B, Rodriguez RL, Huang N (2002) Transgenic Res 11:229–239
222. Wan Y, Lu J, Cui Z (2006) Sep Purif Technol 48:133–142
223. Ireland J, Herzog J, Unanue ER (2006) J Immunol 177:1421–1425
224. Daniels JS, Pourmand N (2007) Electroanalysis 19:1239–1257
225. Lin Y, Allard LF, Sun Y-P (2004) J Phys Chem B 108:3760–3764
226. Shim M, Shi Kam NW, Chen RJ, Li Y, Dai H (2002) Nano Lett 2:285–288
227. Beyer M, Felgenhauer T, Bischoff F, Breitling F, Stadler V (2006) Biomaterials 27:3505–3514
228. Chen ES, Chen ECM (1998) Bioelectrochem Bioenerg 46:15–19
229. Martínez MT, Tseng Y-C, Ormategui N, Loinaz I, Eritja R, Bokor J (2009) Nano Lett 9:530–536

230. Chen RJ, Choi HC, Bangsaruntip S, Yenilmez E, Tang X, Wang Q, Chang Y-L, Dai H (2004) J Am Chem Soc 126:1563–1568
231. Star A, Gabriel J-CP, Bradley K, Grüner G (2003) Nano Lett 3:459–463
232. Lee S-W, Laibinis PE (1998) Biomaterials 19:1669–1675
233. Ruiz-Taylor LA, Martin TL, Zaugg FG, Witte K, Indermuhle P, Nock S, Wagner P (2001) Proc Nat Acad Sci U S A 98:852–857
234. Claesson PM, Blomberg E, Fröberg JC, Nylander T, Arnebrant T (1995) Adv Colloid Interface Sci 57:161–227
235. Wang RLC, Kreuzer HJ, Grunze M (1997) J Phys Chem B 101:9767–9773
236. Kissinger PT, Heineman WR (1983) J Chem Educ 60:702.
237. Steel AB, Herne TM, Tarlov MJ (1998) Anal Chem 70:4670–4677
238. Mansfeld F, Han LT, Lee CC, Chen C, Zhang G, Xiao H (1997) Corros Sci 39:255–279
239. Srisuwan N, Ochoa N, Pébère N, Tribollet B (2008) Corros Sci 50:1245–1250
240. Sagüés AA, Wolan JT, Fex AD, Fawcett TJ (2006) Electrochim Acta 51:1656–1663
241. Sosa E, Cabrera-Sierra R, Oropeza MT, Hernandez F, Casillas N, Tremont R, Cabrera C, Gonzalez I (2003) Electrochim Acta 48:1665–1674
242. Tzvetkov B, Bojinov M, Girginov A, Pébère N (2007) Electrochim Acta 52:7724–7731
243. Nogueira A, Nóvoa XR, Pérez C (2007) Prog Org Coat 59:186–191
244. Wagner N (2005) In: Barsukov E, Macdonald JR (eds) Impedance spectroscopy: theory, experiment and applications. Wiley, New York
245. Manohar AK, Bretschger O, Nealson KH, Mansfeld F (2008) Bioelectrochemistry 72: 149–154
246. Roy SK, Orazem ME (2008) J Power Sour 184:212–219
247. Amstrong RD, Bel MF, Metcalfe AA (1978) Electrochemistry 6:98–127
248. Seland F, Tunold R, Harrington DA (2006) Electrochim Acta 51:3827–3840
249. Kharitonov AB, Alfonta L, Katz E, Willner I (2000) J Electroanal Chem 487:133–141
250. Vladikova D, Raikova G, Stoynov Z, Takenouti H, Kilner J, Skinner S (2005) Solid State Ion 176:2005–2009
251. Kell DB, Davey CL (1990) In: Cass AEG (ed) Biosensors: a practical approach. IRL Press, Oxford
252. Berggren C, Bjarnason B, Johansson G (2001) Electroanalysis 13:173–180
253. Gabrielli C (ed) (1990) Use and application of electrochemical impedance techniques. Solartron analytical, Farnborough
254. Sluyters-Rehbach M, Sluyters JH (eds) (1970) Electroanalytical chemistry. Dekker, New York
255. Katz E, Willner I (2003) Electroanalysis 15:913–947
256. Macdonald JR (ed) (1987) Impedance spectroscopy: emphasizing solid materials and systems. Wiley-Interscience, New York
257. Liu SH (1985) Phys Rev Lett 55:529
258. Heer F, Franks W, Blau A, Taschini S, Ziegler C, Hierlemann A, Baltes H (2004) Biosens Bioelectron 20:358–366
259. Bates JB, Chu YT, Stribling WT (1988) Phys Rev Lett 60:627
260. Kerner Z, Pajkossy T (2000) Electrochim Acta 46:207–211
261. Kong J, Franklin NR, Zhou C, Chapline MG, Peng S, Cho K, Dai H (2000) Science 287:622–625
262. Bradley K, Briman M, Star A, Grüner G (2004) Nano Lett 4:253–256
263. Heller I, Janssens AM, Mannik J, Minot ED, Lemay SG, Dekker C (2007) Nano Lett 8:591–595
264. Maehashi K, Katsura T, Kerman K, Takamura Y, Matsumoto K, Tamiya E (2006) Anal Chem 79:782–787
265. Esplandiu MJ (2005) Contributions Sci 3:33–46
266. Goldstein JI et al. (eds) (1992) Scanning electron microscopy and X-ray microanalysis. Plenum Press, New York
267. Flewitt PEJ, Wild RK (eds) (1994) Physical methods for materials characterization. Institute of Physics Publishing, London

268. Reimer L, Kohl H (eds) (2008) Transmission electron microscopy: physics of image formation. Springer Science, Berlin
269. Zhong Q, Inniss D, Kjoller K, Elings VB (1993) Surf Sci Lett 290:L688–L692
270. Cardiner DJ, Graves PR (eds) (1989) Practical Raman spectroscopy. Springer, Berlin
271. Jorio A, Saito R, Hafner JH, Lieber CM, Hunter M, McClure T, Dresselhaus G, Dresselhaus MS (2001) Phys Rev Lett 86:1118
272. Rafailov PM, Stoll M, Thomsen C (2004) J Phys Chem B 108:19241–19245
273. Anglaret E, Dragin F, Pénicaud A, Martel R (2006) J Phys Chem B 110:3949–3954
274. Mattia D, Rossi MP, Kim BM, Korneva G, Bau HH, Gogotsi Y (2006) J Phys Chem B 110:9850–9855

Chapter 2
Objectives

In all sensor systems, the element or material which makes the determination plays obviously a fundamental role. The introduction of elements with micro/nanometric dimensions is a novel trend to increase the performance of sensors in terms of sensitivity and fast response. Thereby, the general objective of this thesis has been to take advantage of the outstanding electroanalytical properties of carbon nanotubes and to develop new and improved sensing platforms for electrochemical and electronic detection of biorecognition processes.

The starting point has been the group's extensive experience in the immobilization of chemical and biological recognition modifiers to develop graphite-polymer (bio)composites as a transducer matrix. This broad experience has been implemented on the new CNT-based electrodes covering the area from the synthesis of the CNTs to their integration as transducers in the sensing systems for the analytical application in the clinical, nutritional and environmental fields.

We have focused mainly on three different aspects of the carbon nanotube electrochemistry. In each of them, the following specific objectives have pursued:

2.1 To Study the Use of Carbon Nanotubes as Electrochemical Transducers and the Relation Between Their Structure and Their Electrochemical Reactivity

I.(a) Choice and design of different carbon platforms-sensors.

I.(b) Comparison of the transduction capabilities of electrodes based on carbon nanotubes arranged in different configurations with other carbon systems and search for the optimal CNTs and their optimal disposition. Specifically the systems evaluated were: graphite-composite, HOPG and different types of CNT dispersed in a non-conductive polymer matrix (CNT-composite) and high density vertically aligned CNT (carbon nanotube forest electrodes).

M. Pacios Pujadó, *Carbon Nanotubes as Platforms for Biosensors with Electrochemical and Electronic Transduction*, Springer Theses, DOI: 10.1007/978-3-642-31421-6_2,

I.(c) Design and manufacture of nanostructured electrode arrays (UMAs) through opal inversion technique. Achievement of low density of vertically aligned carbon nanotubes grown by plasma enhanced chemical vapour deposition, an optimal configuration of CNTs for enhancing the electrochemical performance and for facilitating (bio)functionalization.

I.(d) Characterization of the micro/nano CNT-electrodes.
- Structural, morphological and electrical study of the different steps in the design of the transducers through nanoscopic and microscopic techniques [Atomic Force Spectroscopy (AFM), Current Sensing Microscopy, Scanning Electron Microscopy (SEM), Energy Dispersion X-ray Spectroscopy (EDS), Transmission Electron Microscopy (TEM)], Raman Spectroscopy and BET isotherm.
- Study of the electrochemical response characteristics of the CNT-transducers using electrochemical techniques such as Cyclic Voltammetry (CV) and Electrochemical Impedance Spectroscopy (EIS). Proof of the usefulness of the EIS technique to detect any electrode surface modification.

2.2 To Study the Use of Carbon Nanotubes as a Suitable Platform for Detecting Biorecognition Events and the Use of Electrochemistry as a Signature of the Carbon Nanotube Functionalization and Sensing

II.(a) To demonstrate that CNTs are a good platform for enhanced electron transfer, that can amplify the signal and that are suitable for (bio)molecular wiring preserving the bioactivity.
- To demonstrate that a vertical disposition of carbon nanotubes is an appropriate configuration for anchorage of biomolecules and has better electrochemical performance.

II.(b) To demonstrate the suitability of the CNTs as electrochemical transducers for enzymatic, DNA-detection and aptamer-based sensors.
- To study and optimize protocols for immobilization of biomolecules on CNTs by X-Ray Photoelectron Spectroscopy (XPS), confocal laser scanning microscopy (CLSM) and Electrochemical Impedance Spectroscopy.

II.(c) To highlight the electroanalytical properties of carbon nanotubes as biosensors.

II.(d) To demonstrate that electrochemical techniques such as Differential Pulse Voltammetry, Cyclic Voltammetry, Chronocoulometry, Electrochemical Impedance Spectroscopy, are sensitive enough and capable to follow the recognition processes.

2.3 To Optimize the Carbon Nanotube-Field Effect Transistor Layout for Real-Time Monitoring of Biorecognition Processes

III.(a) Optimization of Carbon Nanotube-Field Effect Transistor (CNT-FET) for its use in aqueous environment. Passivation of the device.

III.(b) Use of CNT-FET for label-free, real time electrical detection of proteins and as aptasensors.

- To study and optimize of protocols for immobilization of receptors and electrolyte conditions.

Chapter 3
Experimental

In this section we will describe the fabrication process involved in the different carbon-based electrodes and explain how the functionalization for the sensing event is carried out.

Special attention will be paid to carbon nanotube working electrodes, which have been developed in various electrode arrangements. Other carbon systems such as Graphite-Epoxy Composites (GEC) will be also built up as a matter of comparison to evaluate the sensor capabilities of carbon nanotube electrodes.

3.1 Preparation of Carbon Working Electrodes

In this work certain carbon-based electrodes, i.e., rigid carbon (graphite, CNT) composite electrodes, CNT-glassy carbon have been elaborated and evaluated for the electroanalytical utility. Under this context, previous CNT purification and polishing of the electrode surface become key factors for an optimum electroanalytical response.

3.1.1 CNT Purification

One of the most common purification techniques is the "oxidative treatment" of nanotubes by liquid-phase or gas-phase oxidation, which introduce carboxylic groups and some other oxygen-bearing functionalities such as hydroxyl, carbonyl, ester, etc. groups into the tubes.

The aim of this oxidative treatment, as mentioned in Chap. 1, is the oxidative removal of the metallic catalyst particles used in the synthesis of the tubes and of the amorphous carbon, a by product of most synthetic methods. Removal of the catalyst particles is important to minimize the electrochemical interference of these particles with respect to possible electrocatalytical effects. Upon oxidative

M. Pacios Pujadó, *Carbon Nanotubes as Platforms for Biosensors with Electrochemical and Electronic Transduction*, Springer Theses, DOI: 10.1007/978-3-642-31421-6_3,

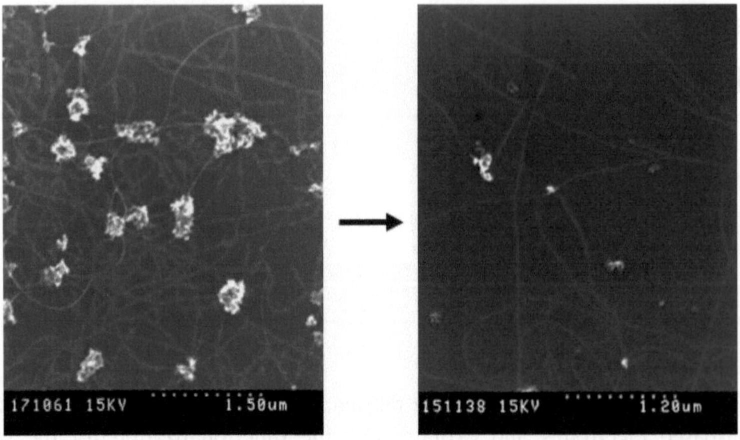

Fig. 3.1 SEM images of SWNT purification: **a** before treatment, **b** after two steps of oxidative purification. CNTs were grown on silica wafers by chemical vapour deposition method

treatment, the introduction of carboxylic groups and other oxygen-bearing groups at the end of the tubes and at defect sites is promoted, decorating the tubes with a somewhat indeterminate number of oxygenated functionalities.

In our case, we treated CNTs before becoming part of the composites or before depositing them on conventional electrodes. Removal of unwanted materials due to CVD growth and the consequently oxidation of the nanotubes was carried out by first heating at 480 °C in presence of oxygen and then immersion in an acidic mixture of 3 M HNO_3/HCl during 2 h (SWNT) or 12 h (MWNT). Specifically, the heat treatment in air helps to remove the amorphous carbon by the oxidation of the carbon layer. After this step, a mixture of HNO_3/HCl dissolves actively the metal nanoparticles at a big extent. The mixture of these two acids is necessary since concentrated HNO_3 can produce a metal oxide layer on the catalyst which would also lead to passivation of the metal nanoparticle and would inactivate the catalyst elimination. In order to avoid that, HCl is added which promotes the dissolution of the oxidized species and avoid the formation of the passivated layer on the metal nanoparticle.

Figure 3.1 shows SEM images of SWNTs prior and after the two steps of oxidative purification (air oxidation followed by HNO_3/HCl treatment). High extent removal of metal catalysts from the CNTs is shown.

3.1.2 Epoxy Composite Electrodes

As mentioned in the introductory part, a composite is a dispersion of a conducting solid material (platinum, gold, graphite, etc.) in an inert matrix, such as a polymer. The first electrochemical transducers based on composites described

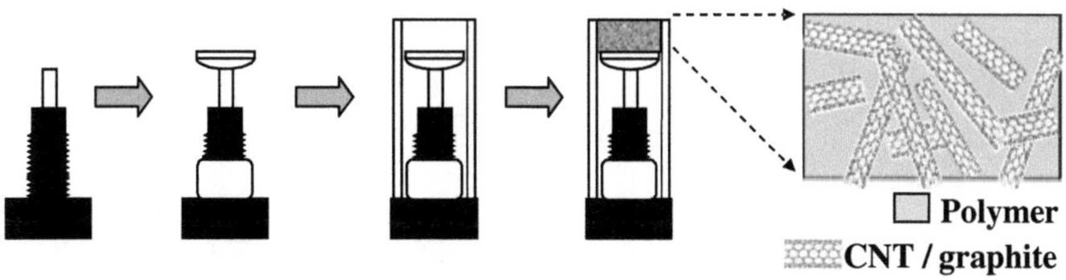

Fig. 3.2 Electrode assembly: electrode body; copper disk; mounting of PVC body; introduction of carbon-epoxy paste

in the literature were the carbon paste, but due to physical and mechanical problems (strength of material, etc.) rigid composites electrodes have been developed, especially those based on epoxy resins [1]. These materials are built by combining a non-conductive epoxy resin (binder matrix) with graphite serving as the conductive material, resulting in the graphite-epoxy composite (GEC).

In this thesis, the construction and evaluation of such electrodes has been carried out together with the novelty of considering the introduction of other forms of carbon in the rigid composites instead of only graphite. Thus, the following lines will briefly describe the build up process of rigid carbon (graphite, CNT) composite electrodes.

Construction of the electrode body [2] is represented in Fig. 3.2. All epoxy composite working electrodes were prepared using a PVC tube body (6 mm i.d.) and a small copper disk soldered at the end of an electrical connector. The conductive paste, different for each case, was deposited filling the cavity in the plastic body. The obtained electrode was then cured in an oven to obtain a solid paste.

The conductive paste of epoxy-carbon was prepared by mixing the two components of the polymer, 80 % (resin and hardener in 20:3 weight ratio) with graphite powder or CNTs at a proportion of 20 % [3]. The homogenization of the reactants was achieved by thoroughly mixing for about 45 min with a spatula. Then, the paste obtained was incorporated into the body of the transducer. The homogenization step is important because the reproducibility of the analysis may be affected by this parameter. The homogenization of carbon nanotubes is more difficult than in the case of graphite because they are much less dense, therefore, requires homogenization times above 45 min. Finally, carbon sensors were cured in an oven at 100 °C during a day.

The thickness of the composites was 3 mm and the contact resistance was in the range of 5–40 ohms, being the graphite composites typically the more resistive ones. It is worth mentioning that the previous acid pre-treatment of the CNT, apart from purifying and generating oxygen functionalities that enhances the electrochemical performance of the composite electrodes, also facilitates a better adhesion between the CNT and the epoxy resin in the composite [4]. Sometimes, strong acidic or electrochemical post-treatments of the electrode composites cannot be carried out due to the lack of stability of the composites to such treatments.

Fig. 3.3 A representative SEM picture of the surface of **a** graphite composite electrode and **b** Carbon nanotube composite

Figure 3.3 shows a representative SEM image of the surface of each type of electrode. From the images, it becomes evident that all these materials share the property of being porous, but have different degrees of porosity. In order to measure the specific surface area of the different composites, we have used the Brunauer–Emmett–Teller method (BET) which is based on the physical adsorption of gas molecules (typically nitrogen) on the solid surface. We have obtained a BET surface area of 1.56 m^2/g for a 30 nm diameter CNT-epoxy composite with a pore volume of 0.006 cm^3/g. In the case of graphite composites, we got a BET surface area of 0.25 m^2/g with a pore volume of 0.003 cm^3/g. From those results, one can observe that the specific surface area of the CNT composite is almost four times bigger and more porous than in the case of the graphite composite.

Electrochemical measurements can also contribute with more insights on those surface features, but this will be discussed later.

We have also performed Raman Spectroscopy to have insights of the structural characteristics of the different electrode systems. Raman analysis of CNT/graphite-epoxy composites, suggest that the epoxy can introduce defects in the nanotube in an extent that can be beneficial for electrochemical studies.

Figure 3.4 shows Raman spectra for high density of vertically aligned CNTs, the CNT-epoxy composite and the graphite-epoxy composite. Specifically, the intensities of the graphite band (G-band) around 1580 cm^{-1} and the disorder band (D-band) around 1350 cm^{-1} were evaluated. The ratio of relative intensities of the D-band to the G-band (I_D/I_G ratio) was extracted as a parameter to determine the structural quality of the carbon. The averaged I_D/I_G ratios for the different electrodes

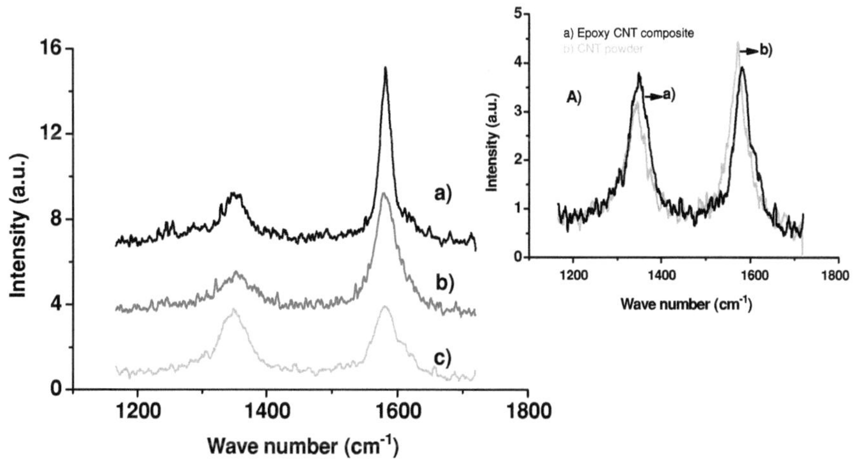

Fig. 3.4 Raman spectra for: **a** the graphite composite, **b** the high density of vertically aligned CNTs electrode and **c** the epoxy CNT composite. *Inset* Raman spectra of the CNT composite and of the purified CNT powder before mixing with the polymer resin

were as follows: 0.64 for graphite composite, 0.72 for the forest-like electrode and 0.96 for the CNT composite. Such a finding indicates that the CNT composite exhibits the highest structural disorder followed by the high density of vertically aligned CNTs electrode and the graphite composite. It was also observed that the polymer resin introduces defects into the CNTs when forming the composite (inset of Fig. 3.4). In the purified CNT powder, the average I_D/I_G ratio was about 0.7 but increases up to an average value of 0.96 when the nanotubes are present in the composite matrix (It is important to stress that the resin alone had no Raman response in the studied wavenumber range.). Therefore, carbon nanotube composites can exhibit not only edge defects on their surface but also additional ones induced by the resin.

3.1.3 CNT-Glassy Carbon

CNTs have been used to modify the surface of a conventional glassy carbon electrode (GC). In our case, we have prepared 1 mg CNT/mL of CNT-casting solutions with dimethylformamide that after gently ultrasonication, were dropped directly onto the renewed surface of the glassy carbon. After drying, the electrode was then ready for use.

The modification of GC with CNTs confers to the electrode some advantages. As it can be seen from Fig. 3.5, the MWNTs-COOH/GC (red solid curve) has larger CV current than bare GC (black dashed curve), which is ascribed to the fact that modification of the electrode with MWNTs-COOH can significantly enhance the effective electrode surface area as well as provide an active binding group for biofunctionalization as we will see later on.

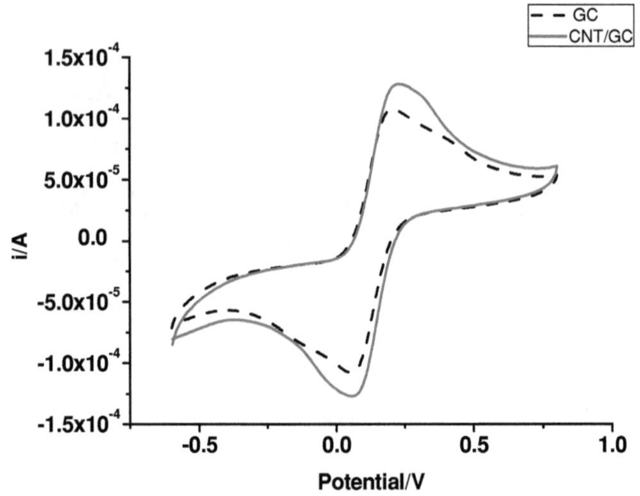

Fig. 3.5 Cyclic voltammograms of the bare glassy carbon electrode (*dashed black line*) and CNT-modified glassy carbon electrode (*solid black line*) in a 2 mM K3Fe(CN)6/K4Fe(CN)6 solution. GC was modified with 5 drops of 3 μl of CNT-DMF solution. Scan rate: 100 mV/s; electrolyte: 1× PBS (pH = 7.4)

3.1.4 Polishing

Electroanalytical response depends on the electron transfer between the electrode material and the species of the solution, and any other type of intermediate layer is of obvious transcendence. Carbon materials are prone to adsorption; this fact makes problematic the preparation of non-contaminated carbon surfaces with adsorbed layers. To prevent poor or irreproducible responses, it is necessary to find conditions where surface impurities have negligible effects or reproducible effects on the process of interest, or prepare the surface of carbon in such a way to avoid the formation of surface layers. Moreover, in some cases, smooth and brilliant surface which is free of physical defects is also required.

Therefore, the renewal of the rigid CNT/graphite composite surface once the electrode is constructed and before running experiments is crucial. The electrode surface was wetted with doubly distilled water and then thoroughly smoothed with different abrasive papers (N° 200-400-600-800-1000-1200) and then with 3 μm alumina paper (polishing strips 301044-001, Orion).

The surface of GC, like other solid electrodes, is also gradually deactivated during exposure to the atmosphere or during the electrochemical use. Therefore, a pretreatment is also needed prior to the CNT modification. The electrode is polished with special polishing cloths with alumina powder containing particles of various sizes. The process begins with the large particles (1 μm) using successively smaller particles until 0.3 or 0.05 μm diameter, verifying the absence of scratches. If 1.0 micron alumina can not remove the scratches, the 1200 grit

Carbimet disk is first used, followed by the sequence of alumina. After polishing with 0.05 micron alumina, the surface of the electrode should be mirror shiny.

3.2 Towards the Design, Fabrication and Optimization of Other CNT Platforms

Most of the electrochemical studies reported in literature are dealing with CNT randomly dispersed onto a substrate electrode or by confining the CNTs on a substrate with polymers. That kind of configuration can be a simple methodology for practical electrochemical studies but it can turn out to be not very suitable for fundamental electrochemical studies For instance, the random configuration of the CNTs essentially causes difficulties in determining the contributions of the ends and the sidewall of the nanotubes to the electrochemical properties of the CNTs. Moreover, such electrodes can also hamper a more controlled (bio)functionalization.

Although the determination of fundamental issues on CNT electrochemistry is a very challenging task, we have put some efforts for moving towards the design, fabrication and optimization of more controlled CNT platforms that could provide us more insights about the CNT electrochemical performance or in the CNT biorecongition process. Accordingly, we have focused our studies in the following platforms: high density vertically aligned CNT macroelectrodes, high density vertically aligned CNT microelectrode arrays, CNT-Ultramicroelectrode Arrays (CNT-UMAs) and Field-Effect Transistors (FET).

3.2.1 High Density Vertically Aligned CNT Macroelectrodes

High density vertically aligned CNT macroelectrodes (or forest CNT macroelectrodes) were used as a matter of comparison with other systems. We have developed such electrodes from commercially aligned catalytic MWNT film grown on silica substrate (MER Corporation), Fig. 3.6.

Figure 3.6b shows the different steps followed for the adaptation of the CNT films for electrodes: the electrical contact was done by placing a silver pad on a corner of the CNT electrode and after that, all the faces of the electrode, except the CNTs, were painted by an insulating ink to avoid any interference in the electrochemical signal. Finally, the insulating ink was also used to define the electrode area on the CNTs face.

3.2.2 High Density Vertically Aligned CNT Microelectrode Arrays

A brief description of the fabrication process of the high density vertically aligned CNT microelectrode arrays will be given. The process takes advantage of the

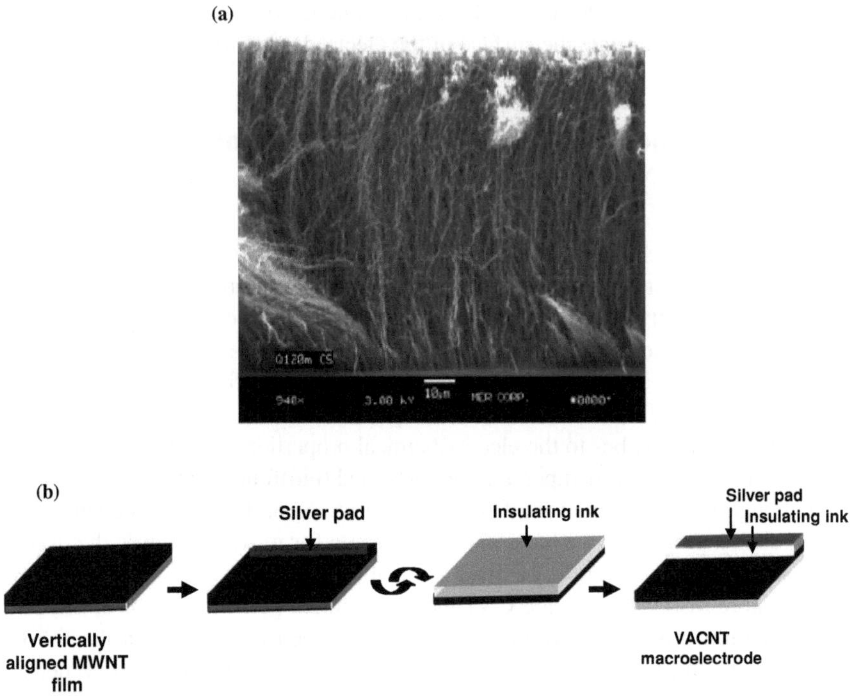

Fig. 3.6 **a** Cross-section SEM image of commercially aligned catalytic MWNT film. Nanotubes are 30–40 nm diameter with 60–80 micron film thickness (nanotube length). **b** Adaptation of the commercial film for its use as a macroelectrode

CNT versatility to develop new miniaturized electrode configurations and was carried out by Dr. Iñigo Martín Fernández at the *Instituto de Microelectrónica de Barcelona* (IBM-CNM, CSIC) [5].

The array of microelectrodes developed consists in four individually addressable electrodes of 300 μm $\times$ 300 μm as depicted in Fig. 3.7. The carbon nanotube growth was performed from a pre-patterned catalyst layer and by performing chemical vapour deposition (CVD). The growth of CNTs from a pre-patterned catalyst layer is the most controllable and uniform strategy to achieve dense and homogeneous CNT layers and to integrate arrays of CNTs on devices [6].

Following, we will outline the novel fabrication process which was used to achieve a high density of vertically aligned CNT layers and to inhibit the catalyst material diffusion to the substrate during the CNT growth.

The starting point was 4 in. p-type Si wafers of low resistivity. A thermal oxide of 1.5 μm was grown on the silicon wafer. Metal electrodes were photolithographically patterned by sputtering a titanium/platinum bilayer (30/150 nm). At this point, it is worthy to mention that titanium is necessary to improve the platinum adherence to the substrate and that Pt acts as metal substrate for the electrode and will also be the catalyst for CNTs growth. CNT synthesis was carried out on

Fig. 3.7 CNT microelectrode with the zoom of two microelectrode patches of 300 μm size each of them. The figure also shows the upright configuration of the CNT growth

platinum because it is one of the most often used materials for electrode and for bio-sensing device fabrication. Platinum is also bio-compatible and it is stable up to very high temperatures.

In order to avoid catalyst diffusion problems and additionally enhance smaller catalyst particle formation for guaranteeing good CNT growth, an additional thin platinum layer was deposited and separated from the thick Pt layer acting as metal electrode [7]. Therefore, a 15 nm thick silicon oxide layer was deposited by Plasma Enhanced Chemical Vapour Deposition (PECVD). After that, a 4 nm of Pt layer was sputtered on top of the SiO_2. Thus, the thin SiO_2 layer between the two platinum ones avoids the diffusion of Pt forming the electrode (thick Pt layer) and the platinum acting as a catalyst (thin Pt layer).

Selective growth of the CNTs was then performed by Rapid Thermal CVD in two step process: the catalyst activation step (increasing the chamber temperature up to 800 °C in 120 s and maintaining that temperature for 180 s in hydrogen atmosphere, 500 sccm) and CNT growth step (800 °C with methane, 1000 sccm for 5 min).

The final fabrication step consisted in introducing the devices in a HF based solution to etch the 15 nm SiO_2 layer in the contact pads after the wafers had been diced. In this way, one would optimise the contact between the electrode and the MWNTs to assure correct wire-bonding at the connection pads.

Figure 3.8 shows a simplified scheme of the fabrication steps of the CNT microelectrode array.

3.2.3 CNT-Ultramicroelectrode Arrays (CNT-UMAs)

In this section, we will describe the first steps of fabrication and design of carbon nanotube-based ordered arrays of nanostructured electrodes by using self-assembly moulds ("opal inversion" technique). As mentioned in Chap. 1, these CNT-UMAs are expected to exhibit exceptional properties for electroanalytical purposes and have been designed with innovative technologies and novel materials which allow us to get well-defined and well-ordered micro/nanostructures on a

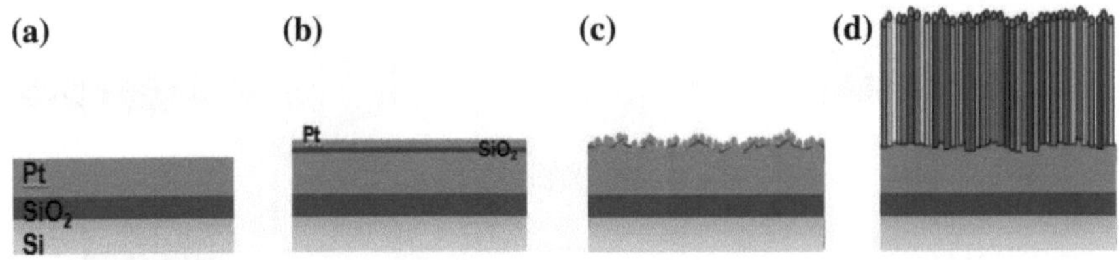

Fig. 3.8 Scheme showing the fabrication process: **a** patterning of the 150 nm thick Pt layer, **b** deposition of the 15 nm SiO_2 layer followed by a 4 nm Pt layer, **c** cracking of the thin Pt layer with temperature, **d** CNT growth and etching of the thin SiO_2 layer

Fig. 3.9 Different steps in the nanostructure generation on a surface by the "opal inversion" technique

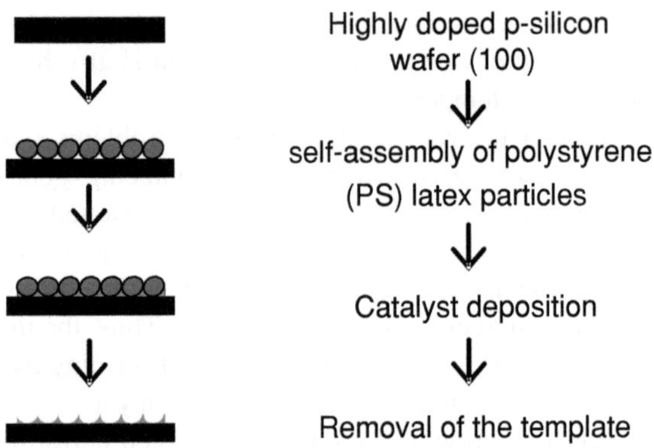

surface. We resort to simple skills and to low cost chemical methods combining nanosphere lithography to control the site density of CNTs and plasma enhanced chemical vapour deposition (PECVD) to grow the CNTs. In the next future, these UMAs could represent the basis for the design of (bio)sensors for which, different (bio)molecule immobilization techniques could be performed.

For the preparation and production of this transducers based on arrays of ultramicroelectrodes, different methodologies are used. The one proposed is based on the concept of Self-Assembly of an inert template and on the deposition of the CNT catalyst by different physicochemical processes. The technique is relatively simple and of great accessibility that takes advantage of the properties of capacity of self organization of some materials to form highly ordered structures on surfaces. As an example of these types of materials, we can mention polystyrene microspheres, which dispersed on a substrate, tend to form structures with hexagonal packaging [8]. These materials turn out to be very versatile as templates for the formation of porous patterns or highly periodic micro/nanostructures by means of the technique "opal inversion".

Figure 3.9 shows the scheme of the "opal inversion" procedure. The technique consists in the self-assembly of the spheres, followed by the deposition of the catalytic material that can be accomplished by different ways: by

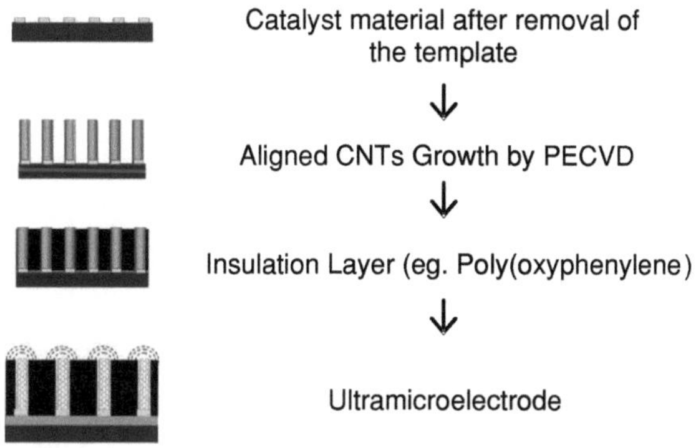

Fig. 3.10 Different processes involved in the design of the UMAs based on carbon nanotubes

electrochemical deposition, electroless deposition, chemical precipitation, PECVD sputtering, etc. The deposition of the material is carried out through the interstitial spaces that leave the hexagonal packaging of the spheres. Subsequently, the spheres of polystyrene are eliminated with blue tape plus dichloromethane (DCM) remaining micro/nanostructures of the deposited material. The size of the micro/nanostructures can be variable from the micro to nano range depending on the size of the particles used as a template.

To enjoy the benefits of the UMA it would be of great significance to generate vertically aligned carbon nanotubes of low density or dispersed as patches on an inert matrix. In order to accomplish that, it is very important to obtain well-ordered metallic spots of the catalyst with a controlled interspacing which will, further on, determine the interspacing of the nanotubes. Figure 3.10 schematizes the final steps to obtain the CNT-UMAS.

All the details of the UMA optimization process will be given and discussed below; it consists mainly of the just mentioned five steps: control of the self-Assembly processes of the microsphere mould, control of the deposition of catalyst material through the mould of microspheres, development and control of the elimination process of the mould, control of the carbon nanotube growth and development and control of encapsulating processes of the ultramicroelectrodes.

3.2.3.1 Control of the Self-Assembly Processes of the Microsphere Mould

Self-assembly of the hexagonal closed-packed monolayer of latex spheres, is the basis of the nanosphere lithography (NSL) [9–11]. This technique is used for creation of masks by deposition of various materials, typically through electrochemical, e-electron beam and sputtering deposition. It is known that NSL can be used to make honeycomb lattices of triangularly shaped islands on various substrates [12]. For accomplishing that, we have employed commercially available monodisperse suspensions of polystyrene nanospheres. Using spheres with different diameters, one can change the spacing and size of the periodically arranged islands.

Fig. 3.11 **a** Diluted solution of PS; **b** solution without surfactant; **c** monolayer assembly in presence of divalent ions

The starting substrates were 1 × 1 cm^2 low resistive highly doped (n-type) which contain arsenic as a dopant, with a native oxide layer. Silicon substrates were cleaned with acetone, isopropanol and Milli-Q water soaks and dried in a nitrogen stream. Alternatively, 100 nm of Pt was deposited making the substrates highly conductive. 5 nm of chromium was first deposited by sputtering in order to improve the platinum conductive layer adherence to the Si substrates.

The first purpose to achieve was the control of the Self-Assembly process of the polystyrene microsphere template in order to form a well-ordered and compact self-assembled monolayer on the conductive substrates. Accordingly, two different approaches were used.

In the first one, the naked substrates (without platinum deposition) were immersed in a liquid dispersion of the microspheres. Subsequently, the substrates were subjected to controlled evaporation of the liquid. The method of evaporation was by angle evaporation: 1 cm diameter cap was filled with a small amount of PS dispersion (140 μL) and the substrate was placed inside supported by the walls of the cap forming an angle of 90°. The particles move during the early stages of evaporation by Brownian motion; then, the attractive capillarity forces of the colloidal spheres induce the formation of compact hexagonal layers. To obtain a homogeneous monolayer and to avoid the agglomeration of particles, it is necessary that the spheres have good interaction with the surface of the substrate. In this sense, it was essential the addition of surfactant agents (dodecylsulfate) to the microsphere liquid dispersion. The substrate, which is slightly negatively charged, was also pre-treated with magnesium salt in order to charge it positively and give fine interaction with the negatively charged particles.

The quantity of deposited spheres, the sphere size (e.g. 1 μm, 900 nm, 600 nm, etc.) was also adjusted. For instance, Fig. 3.11 shows different conditions and the resulting assembly of the microspheres: a low concentration of them results in a poor coverage of the substrate; the absence of surfactant causes an agglomeration of the spheres; however, with the right pre-treatment of the substrate and a

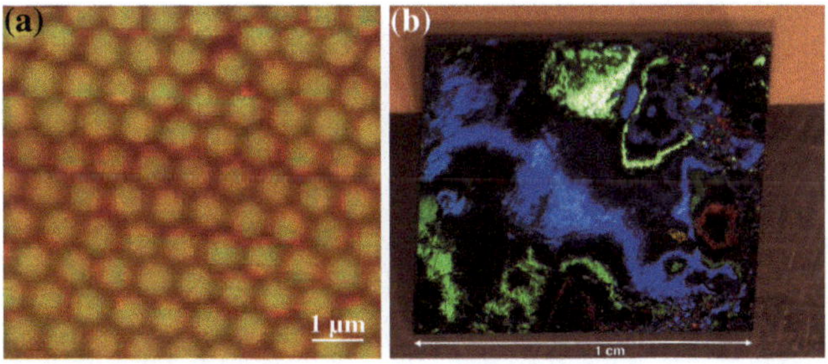

Fig. 3.12 Highly ordered monolayer of polystyrene nanospheres (diameter 0.8 μm) on Si substrate made by the self-assembly nanosphere lithography. **a** AFM image of the *closely packed hexagonal* nanospheres, **b** diffraction colour image showing a big area of diffraction in the *blue region*

control of the mentioned parameters, results in the desired monolayer assembly of polystyrene latex particles.

We have followed a second approach which gave even better results. This approach was carried out in collaboration with Guillaume Sauthier, Ph.D. and technician of the PLD & Nanoionics Group of CIN2. In this case, the highly conductive substrate $Si/SiO_2/Pt$ was used for the deposition of polystyrene (PS) latex particles. In this first step, the latex particles (diameter of 800 nm) were slowly applied to the interface of an ethanol/water (1:1) solution contained in a 15-cm Petri dish by using a glass pipette. The amount of solution that could be distributed to cover the whole aqueous surface with a hexagonal close packed monolayer was limited by the size of the Petri dish and the diameter of the spheres. About 70 % of the particle solution was applied to the surface of the water, which left some places for stress relaxation and avoided the formation of cracks in the lattice during the next steps of the preparation. At this stage the monolayer had the biggest domains of about 2 cm^2 with very irregular shapes. To promote the growth of large domains, gentle waves were applied to the liquid medium by slow and careful vessel tilting. A drop of Triton X-100 diluted with methanol by 1:400 was added to get more packed structures. After this treatment, large crystals that showed clear diffraction colours were created (Fig. 3.12). Finally, the monolayer was transferred on the substrate by carefully submerging the substrate under the packed structure and then lifting it nearly horizontally to pick up the dense microsphere packings on its upper side. The deposited monolayer was then dried in a clean atmosphere. The diffraction colour in Fig. 3.12b shows that indeed we got a highly ordered monolayer.

3.2.3.2 Catalyst Deposition Through the Mould and Mould Removal

Different approaches can be used for thin film catalyst deposition: electrochemical deposition [13], PECVD sputtering, electroless deposition, electron beam or thermal evaporation, etc. Catalyst pattern formation in the interstices of the

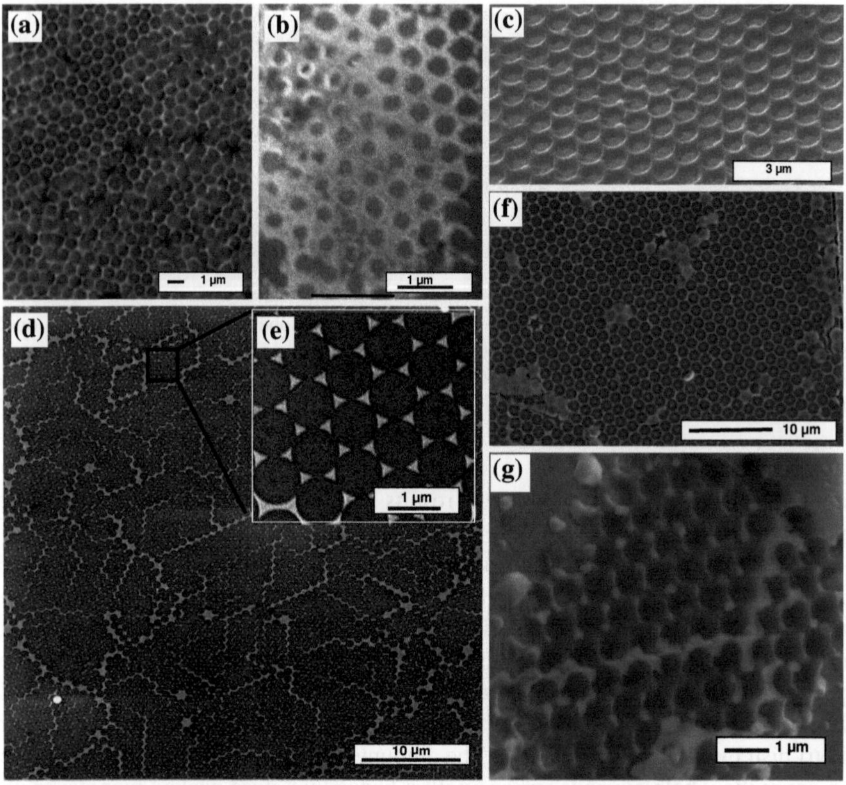

Fig. 3.13 SEM images of different methods of catalyst/metal deposition; **a** electrochemical deposition of Cu; **b** PECVD Sputtering of Fe; **c** evaporation of Fe; **d** evaporation of Au showing the big domain reached after removal of PS and **e** zoom of the honeycomb pattern of Au; **f** evaporation of Co; **g** electrochemical deposition of Fe

nanosphere template, for the later on CNT growth, is presented as an alternative to other catalyst deposition techniques such as dots made by electron beam (e-beam) lithography that are very expensive and limited to a very small area.

The nanosphere templates were subsequently used as a mask for deposition of the catalyst (Fe, Co, Ni, etc.) with the different methods. In this manner, a well-ordered structure of spots could be obtained whose interspacing will be determined and controlled by the size of the spheres forming the template, by the quantity of the deposited catalyst material and also by thermal post-treatment. Figure 3.13 shows patterns achieved with different methods of deposition of various metals.

In a final approach, PECVD evaporation of Fe/Ni catalyst was also employed. The thickness of the catalyst film can range from few nanometres to tens of nanometres. We evaporated 4 nm of catalyst. It is worth keeping in mind that the diameter of the nanotubes will also be determined by the catalyst film thickness and that the type of CNT grown will be determined by the kind of catalyst used.

After the deposition of the catalyst, the polystyrene particles can be removed either chemically with dichloroethane or mechanically by blue tape followed by soaking the substrates in dichloromethane for 10 min. The process with blue tape gave better results.

The resulting pattern can be either a connecting nanostructurated pattern (Fig. 3.13a–d) or a non-connecting pattern of the catalyst (Fig. 3.13e, f). The metal nature was verified by Energy-dispersive X-ray spectroscopy.

During CNT growth and due to the elevated temperature, the connecting catalyst pattern breaks forming separated nanoparticles or in the case of non-connecting nanostructure, the metal patches forming the hexagons acquire a more rounded shape (drop-like shape). All these effects are due to temperature induced minimization of surface tension. From these more rounded nanostructures, the CNTs start nucleation and growth.

3.2.3.3 Carbon Nanotube Growth

Taking as base the substrates modified with the catalyst, the growth of the carbon nanotubes was carried out by plasma enhanced chemical vapour deposition (PECVD) [14]. Among carbon nanotube production techniques, chemical vapour deposition is one of the most widely used, specially for applications in which a high control over length, diameter and positioning of the CNTs is desired [15]. Particularly, we used plasma enhanced CVD because favours an excellent vertically aligned growth of the CNTs guided by the direction of the electric field established in the plasma sheath [16].

The parameters of the chemical vapour deposition (temperature, pressure, gas flow, deposition time, plasma intensity) were varied to define the influence of such parameters in the characteristics of the resultant nanotubes.

Especially, we aimed at controlling the length and orientation of the nanotubes, while their diameter and density were better defined by the catalyst material prepared in the previous step (Fig. 3.14).

The catalyst-coated was transferred in air into the chamber and heated in H_2 up to 750 °C at a rate of 40 °C/min. Once the working temperature was stabilized, a 50 sccm flow of NH_3 was entered in the chamber replacing the existing H_2. After H_2 was removed and pressure got stable, a plasma was ignited and after 2 min, a 25 sccm C_2H_2 flow was entered. The waiting time between plasma ignition and C_2H_2 flow entrance allowed the achievement of stable discharge conditions. The working pressure was set constant at 100 Pa.

In such PECVD procedure, acetylene gas (C_2H_2) was used as the carbon source, whereas ammonia gas (NH_3) was used as plasma enhancer and growth promoter.

3.2.3.4 Development and Control of Encapsulating Processes of the Ultramicroelectrodes

Different techniques can be explored to passivate the free spaces between nanotubes in order to confer greater performance as ultramicroelectrode arrays and

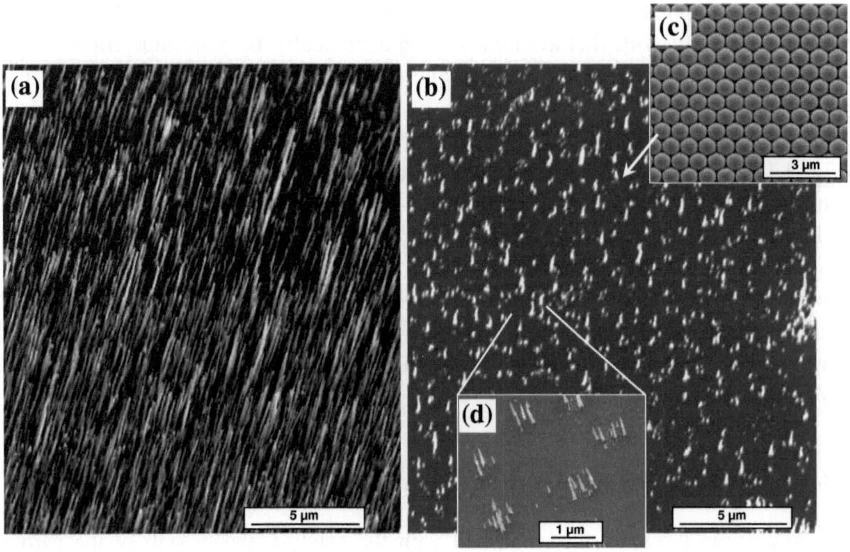

Fig. 3.14 SEM images **a** tilted image of high density growth of CNTs from a non-patterned surface; **b** low density of CNTs from patterned surface; **c** template for the CNT fabrication; **d** zoom of highly periodic vertically aligned CNTs

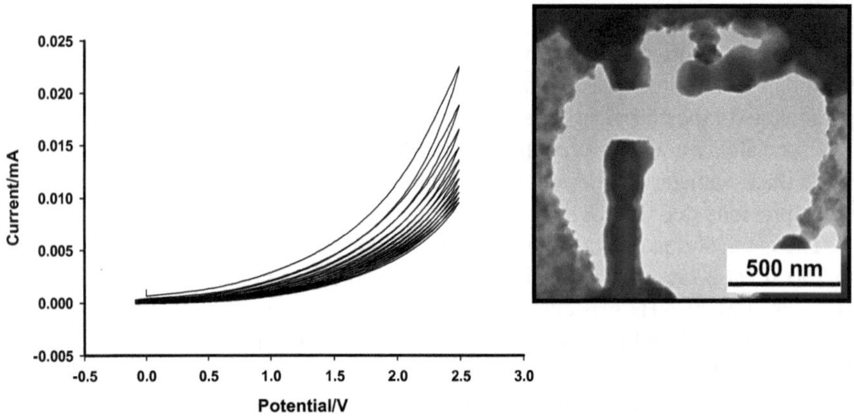

Fig. 3.15 Passivation by the electropolymerization of Poly(oxyphenylene) film. TEM images showing encapsulation of the CNTs

to increase the robustness of the carbon nanotube system. We have tested the electropolymerization of non-conductive polymers. Specifically, we have electropolymerized polyoxyphenylene from phenol monomers in the following conditions: 30 ml solution 1:1 Methanol:water, 60 mM phenol, 90 mM allylphenol, 160 mM butoxyethanol; pH 9,5, 30 min at 2.5 V). Under these conditions, a polymer layer of 60 nm was deposited. Such layers exhibited high resistivities of

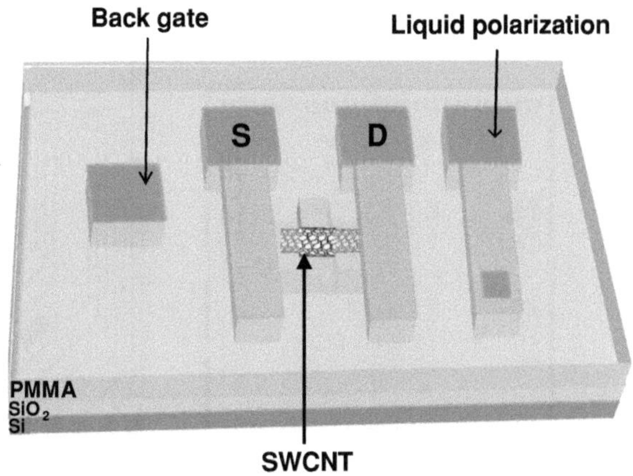

Fig. 3.16 CNT-FET structure formed of a SWNT contacted by two metal electrodes and actuated by a back-gate. Liquid polarization electrode is included in order to work in liquid environment. The sensor is fully passivated except on the CNT channel region, on the liquid polarization electrode and on the connection pads

about $2.8 \times 10^{11}\ \Omega$ m and low dielectric constants as extracted from Impedance Spectroscopy. Figure 3.15 shows the electropolymerizacion process on the CNT patterned electrode. It can be observed a decrease of current with time as an insulating polymer layer is generated on the CNT electrode. The encapsulation is an important process since helps to reduce the background noise and to enlarge the sensitivity of the electrodes during the measurement.

In the case that the tip terminals of the nanotubes resulted covered, we applied different methodologies to eliminate the insulating material on the tips of the nanotubes as potential pulses (about 2 V in KCl solutions) or soft plasma etching (200 W for 12 s).

3.2.4 Carbon Nanotube Field-Effect Transistor (CNT-FET)

It is important to highlight that the design and fabrication of CNT-FET sensors were carried out by Dr. Iñigo Martín Fernández at the *Instituto de Microelectrónica de Barcelona* (IBM-CNM, CSIC) [6]. However, as we have used this sensor configuration for biorecognition processes, a brief description of the development of such system is needed in order to better understand the subsequent biosensing mechanism.

A simple scheme of the final CNT-FET sensor is shown in Fig. 3.16.

The fabrication of the passivated CNT-FET sensor is divided into two main steps: the fabrication of the CNT-FET structures and their passivation process. In all the steps, standard lithographic steps have been used (photolithography, lift-off process, plasma etching, electron beam lithography, etc.)

3.2.4.1 Technological Process of the CNTs-FET

The main aim was to develop a technological process to achieve fabrication of a large number of functional CNT-FETs at wafer level fabricated by means of

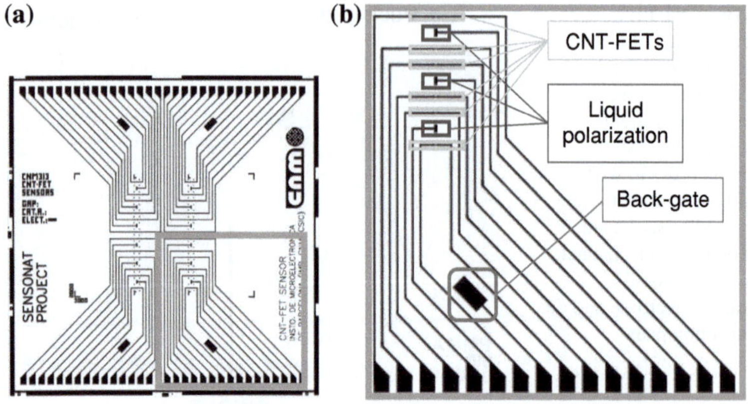

Fig. 3.17 **a** Schematic of 1×1 cm^2 chip platform. **b** Elements on a quadrant: 6 CNT-FET structures, back-gate electrode and 3 liquid polarization electrodes and 16 pad matrix

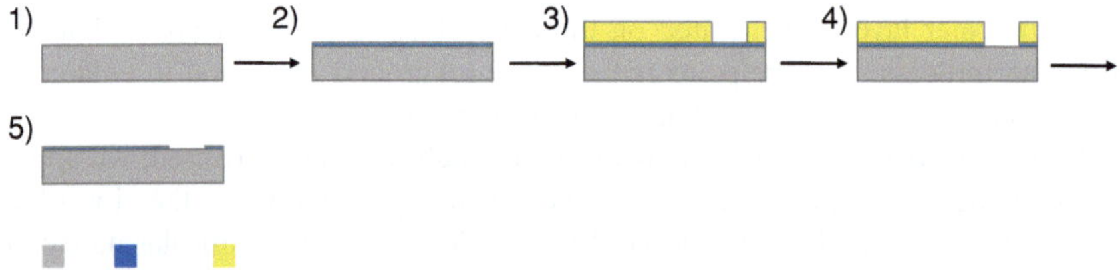

Fig. 3.18 Substrate preparation. From 1) 4″ wafer to 5) the formation of the back-gate contact. 2) Thermal oxidation, 3) photolithography, 4) RIE of silicon oxide

conventional microelectronic steps (neither E-beam lithography nor AFM was needed, just photolithography as a patterning technique).

The first step before the transistor fabrication was the design of the CNT-FET sensing chip platform, Fig. 3.17. The platform design aimed at concentrating the maximum number of CNT-FET structures at the centre of the chip, far from the connection pads. The platform consisted of four quadrants symmetrically identical.

The fabrication process which included three photolithographic levels and 18 process steps was divided in three parts: the substrate preparation, the selective synthesis of the SWNTs and the electrode patterning.

In the substrate preparation, Fig. 3.18, a 4 in., 500 μm thick and highly p-doped silicon wafer was further doped with phosphorous to optimise the gate actuation in the CNT-FET. Then, a 100 nm thick silicon oxide layer was thermally grown in dry oxygen atmosphere at 1.100 °C. After these first steps, photolithography was carried out to allow patterning the windows to connect the bulk material and the metal electrodes to form the back-gate contact. Finally silicon oxide was etched by Reactive Ion Etching (RIE) and the photoresist was removed by oxygen plasma.

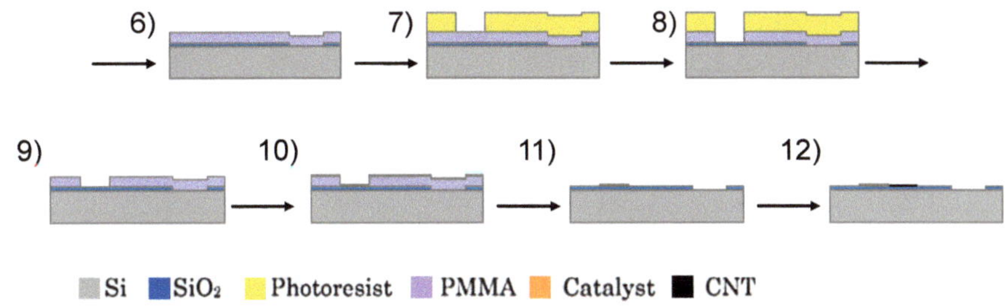

Fig. 3.19 Selective synthesis of the SWNTs. 6) PMMA deposition, 7) photolithography, 8) patterning of PMMA, 9) photoresist removal, 10) catalyst deposition, 11) PMMA lift-off, 12) chemical vapour deposition

The selective synthesis of the CNTs was accomplished by the selective deposition of the CNT catalyst material, Fig. 3.19. First a 35 K molecular weight polymethyl methacrylate (PMMA) layer was deposited by spin coating. Then, a photoresist layer was deposited on top of the PMMA and photolithography was carried out in order to define the areas on which the catalyst will be deposited. After that, PMMA was also patterned by oxygen plasma just at the areas below of the irradiated photoresin. The wafers are then immersed in isopropyl alcohol to remove the photoresist, but not the PMMA layer and rinsed with de-ionised water. The final step before the growth of the carbon nanotubes was to deposit the catalyst material by spin coating. Ferric nitrate and molybdenum embedded alumina particles, $Fe/Mo/Al_2O_3$, was a complex catalyst that is often used in literature for SWNT synthesis [17] and therefore was the one chosen for the selective synthesis of the CNTs. The lift-off of the PMMA-catalyst layer was then performed in acetone. Low density of SWNTs were synthesized employing rapid thermal CVD (RTCVD): first an activation of the catalyst was needed by increasing the chamber temperature up to 800 °C in two minutes and maintaining that temperature for another three minutes at 400 sccm H_2; after that, the growth of carbon nanotubes was carried out at 800 °C, 1200 sccm CH_4 and 400 sccm H_2 during 6 min.

The synthesised SWNTs were between 1 and 10 μm long and with diameters about 2 nm.

The third main issue of the process was electrode patterning on the CNTs, Fig. 3.20. First, photolithography was used to pattern the contact metals, tracks and pads. A 10 nm Ti and 100 nm Au was deposited afterwards by sputtering. Finally, the metal lift-off was performed in acetone.

3.2.4.2 Passivated CNT-FET Sensors for Electronic Detection in Liquid Environment

As mentioned in Chap. 1 one of the interesting applications of the FET is the real-time monitoring of (bio)processes in aqueous environment. That is a topic that has been

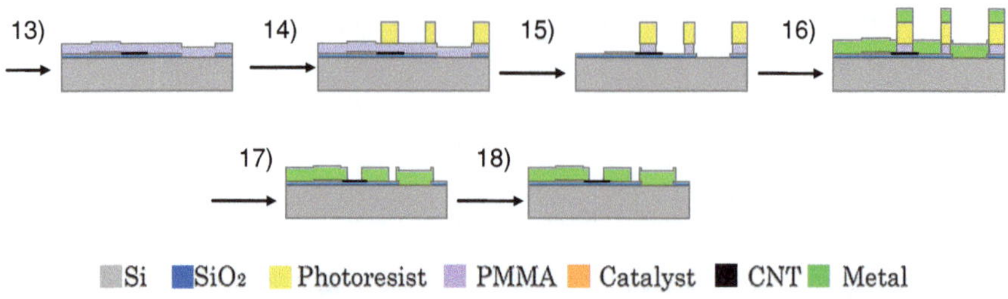

Fig. 3.20 Electrode patterning. 13) Deposition PMMA, 14) photolithography, 15) patterning of PMMA with RIE, 16) metal deposition, 17) metal lift-off, 18) wafer dicing

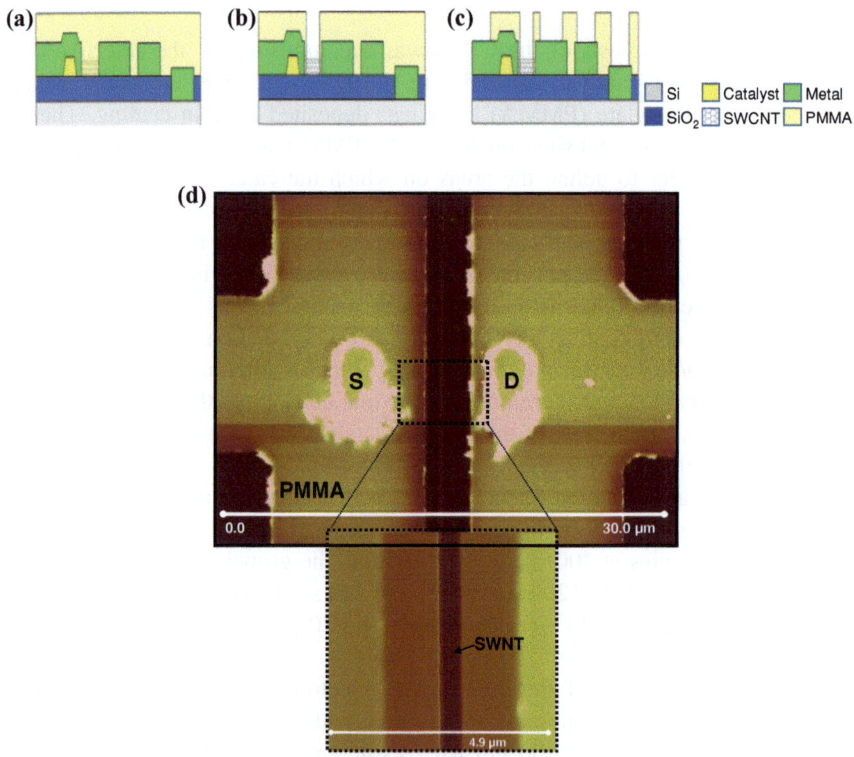

Fig. 3.21 Steps of the passivation procedure. **a** PMMA coating, **b** e-beam lithography, **c** removal of the PMMA from the pads, **d** AFM image of the final passivated CNT-FET device

scarcely explored in literature since most of the CNT-FET biosensing studies have been performed in dried conditions. However and in order to have reliable results in aqueous environment, the passivation of the metal contacts is mandatory to avoid electrochemical side reactions that can degrade the electrodes or smear the conductance signal from the CNT channel. Additionally, E-beam lithography was selected as the lithographic technique that defined the non-passivated regions of the CNT-FET device. This precise technique was necessary because of the major difficulty of the alignment of the pattern with respect to the already fabricated structures.

The procedure represented in Fig. 3.21, started by coating the chips with a 100 nm thick 950 K molecular weight PMMA layer assuring the coverage of all the structures. E-beam lithography was then employed to pattern the passivation layer on the channels of the CNT-FETs and on the liquid polarization electrodes. Finally, the PMMA was removed from the pads with a swap soaked with acetone. A scheme of the final passivated electrodes is also shown in Fig. 3.21d, where a zoom of the cannel of the CNT-FET is depicted. Note that the SWNT is not visible because of the big differences of height as compared to the CNT height.

3.3 Procedures for CNT-Functionalization

CNT-functionalization is one of the prerequisites to be firstly established for the development of applications with such CNT nanostructures.

Functionalization allows immobilization of biomolecules on CNTs and this motivates the use of nanotubes as potentially new types of biosensor materials.

In this section, we will deal with the procedures for electrode modification in order to anchor covalent and non-covalent biorecognition elements in our sensing systems. Such functionalization protocols were used to immobilize redox proteins or enzymes or DNA. In the case of redox proteins/enzymes we have pursued two important aspects: the achievement of direct electron transfer between the electrodes and the redox active (bio)molecules and the preservation at the same time of their bioactivity.

3.3.1 Electrode Modification

Oxygen moieties are necessary for covalent attachment by carbodiimide chemistry, as previously discussed. Thus, electrochemical pre-treatments were performed on CNT microelectrodes and on forest CNT electrodes for cleaning the CNTs by oxidizing the amorphous carbon and for generating carboxylic moieties for the later covalent functionalization. On CNT/Glassy carbon and carbon nanotube composites, the carboxylic moieties were generated with the oxidative purification of the nanotubes before becoming part of the electrodes.

The different electrochemical pretreatments were: 1 M HNO_3 electrolyte at 1 V for 10 s, 0.1 M KCl at 1.75 V for 60 s and 1 M NaOH at 1 V for 60 s. Figure 3.22 shows that treatments in NaOH solutions provided the best electrochemical response based on cylic voltammetry measurements (CV) in presence of a redox

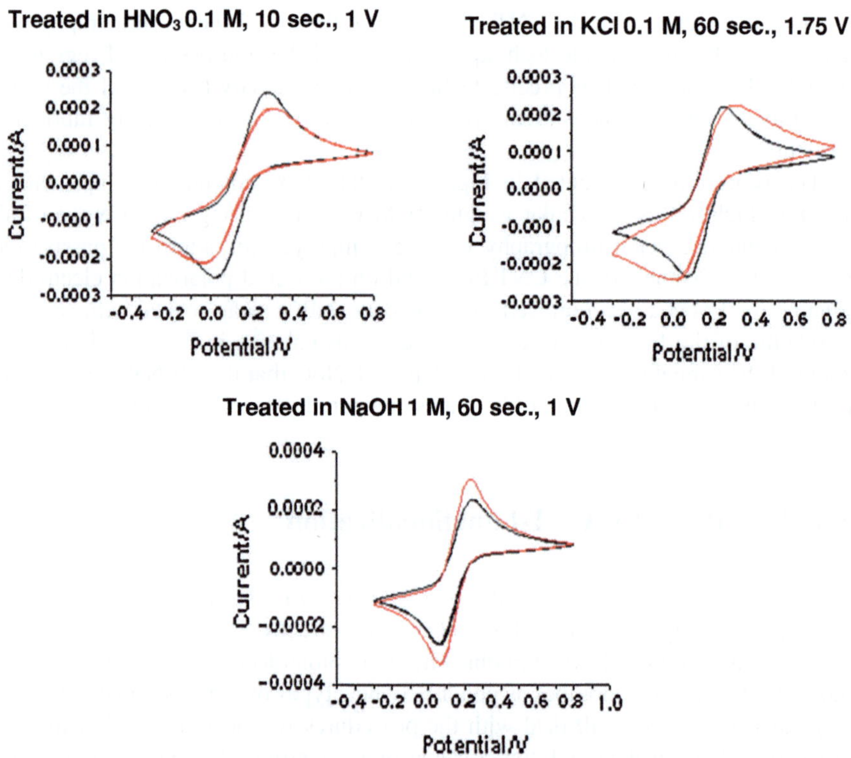

Fig. 3.22 Electrochemical pre-treatment of micro electrodes. *Black* before treatment and *red* after treatment (in K3[Fe(CN)6]/K4[Fe(CN) 6] (1:1)) ΔEp decreases in 33 mV when the treatment is performed with NaOH. In the other cases, ΔEp increases

benchmark $Fe(CN)_6^{3-/4-}$ by using cyclic voltammetry. Specifically, we observed for such treatment higher current signal and a decrease in the peak potential difference as indication of the enhancement of the electrochemical performance (increase of the electrode reversibility). In the case of the forest CNT electrodes, also a heat treatment (at 470 °C) in air for an hour was performed.

3.3.1.1 Covalent Attachment

In the covalent attachment, molecules are directly attached to the CNTs via the formation of chemical bonds. Graphitic sidewall functionalization will unavoidably lead to some modifications of the nanotube delocalized π-system, however we have followed such protocol on MWNTs in which some perturbation is not detrimental as compared in the case of SWNTs.

One of the most common schemes for the covalent functionalization is by means of the carbodiimide chemistry, which consists, as stated, on the generation of carboxylic groups at the edges of the CNT (electrode) and then the coupling with NH_2 containing species (biomolecules) through the formation of a covalent amide bond [18]. That procedure has been extensively used for the attachment of amine terminated DNA aptamers, enzymes and proteins. The strategies that were followed will be discussed next.

Covalent Attachement of NH_2-Oligonucleotides for DNA-Detection

The first step of the immobilization procedure consisted in the activation of the –COOH terminal groups of the CNTs of the electrodes by immersing them in 0.05 M EDC/0.03 M Sulfo-NHS in PBS solution (0.1 M NaCl + 0.01 M sodium phosphate buffer, pH = 7) for 15 min. The electrodes were then thoroughly rinsed with PBS. To carry out ss-DNA probe immobilization, electrodes were incubated overnight in aminated DNA solution (e.g. 1 μM Poly-C-NH_2 in PBS) at room temperature under a wet environment. This was followed by a washing step with hydroxylamine/SDS (by immersing in 0.05 % SDS + 0.04 M Hydroxylamine solution) for 15 min to remove non-specific adsorption and to deactivate free –COOH groups.

Before hybridization a blocking step was performed in order to avoid non-specific adsorption of target oligonucleotide. The electrode surface was treated with a 0.01 M solution of PEG in PBS, pH = 7 for 15 min and then rinsed with PBS. Hybridization was performed incubating the NH_2-probe modified electrodes with different concentrations of the complementary target strand (from 10^{-10} to 10^{-6} M in PBS buffer) of Poly-G (20 bases). The hybridization was achieved at room temperature, for 30 min. This was followed by a gentle washing step in PBS buffer. Negative controls were also made with non-complementary oligonucleotide (Poly-T).

It is worth highlighting that this strategy is interesting because covalently modified DNA probes exhibit structural flexibility and chemical stability, thus improving hybridization efficiency [19–21].

Covalent Attachment of Protein/Enzymes for Catalysis

The strategy followed with proteins is similar than the one of DNA. Again myoglobin and catalase were attached onto the ends of the MWNT by using water soluble carbodiimide, EDC, to promote amide linkages between the carboxyl-terminated nanotubes and the lysine residue of the proteins. In this case, we used one step approach; this means immersing the carbon electrodes in Mb or Cat solutions in PBS (3 mg/ml) together with EDC (3 mg/ml) during 24 h. Finally, electrodes were thoroughly washed in PBS for one hour and dried with nitrogen. In the case of other electrode substrates such as graphite, gold and HOPG, the electrodes were only immersed in protein solutions in PBS (3 mg/ml) for 24 h to promote its adsorption and cleaned in the same way as mentioned above.

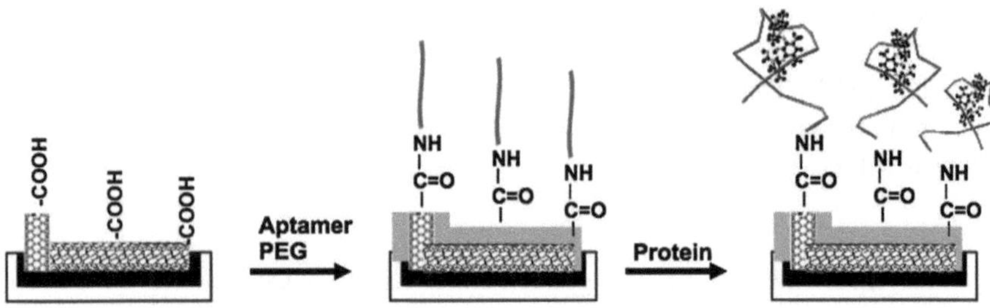

Fig. 3.23 Protein detection strategy using aptamers. Amide bond between CNT-COOH and NH2-aptamer is achieved via carbodiimide chemistry. Polymer PEG is used to block non-specific adsorption of proteins on the CNT

Covalent Attachment of DNA for Protein Sensing

For aptamer attachment, the same approach of carbodiimide chemistry was used. To achieve that, aptamers were purchased modified at the 5′-terminus with a primary aliphatic amine based linker.

First we had to assure proper folding of the aptamer and break any pre-existent structure. For doing that we have performed a denaturation step heating the solution of aptamer (1.0 μM in PBS) to 90 °C for 3 min and then cooling by immersion into ice bath.

After that, the cleaned and activated working carbon electrode was immersed in 200 μl of the aminated-aptamer solution containing EDC/NHS 1 mg/0.5 mg, and incubated overnight. After that, a washing step with PBS buffer was carried out in order to remove the leftover reactives and then a blocking step was performed in order to avoid non-specific adsorption of proteins. The electrode surface was treated with a 0.04 M solution of PEG or BSA in PBS for 15 min and then rinsed with PBS. The resulting modified aptamer electrode was incubated 15 min with the protein analyte (proteins thrombin and lysozyme were studied in the range from 1.0 μM to 1.0 pM) and washed with PBS.

Control experiments were performed by treating the aptamer-functionalizated electrodes with non-specific target, 1.0 μM elastase or myoglobin (in the case of thrombin) and 1.0 μM cytochrome C (in the case of lysozyme) for the same period of time.

In Fig. 3.23 we can see the scheme of the functionalization strategy using aptamers for protein recognition.

3.3.1.2 Non-covalent Attachment

The non-covalent functionalization relies on physical adsorption that is based on weak interactions (e.g. hydrogen bonding, π–π stacking, electrostatic forces, Van der Waals forces and hydrophobic interactions). It is particularly attractive because it offers the possibility of attaching chemical handles preserving the sp [2] nanotube structure and consequently not affecting the electronic network.

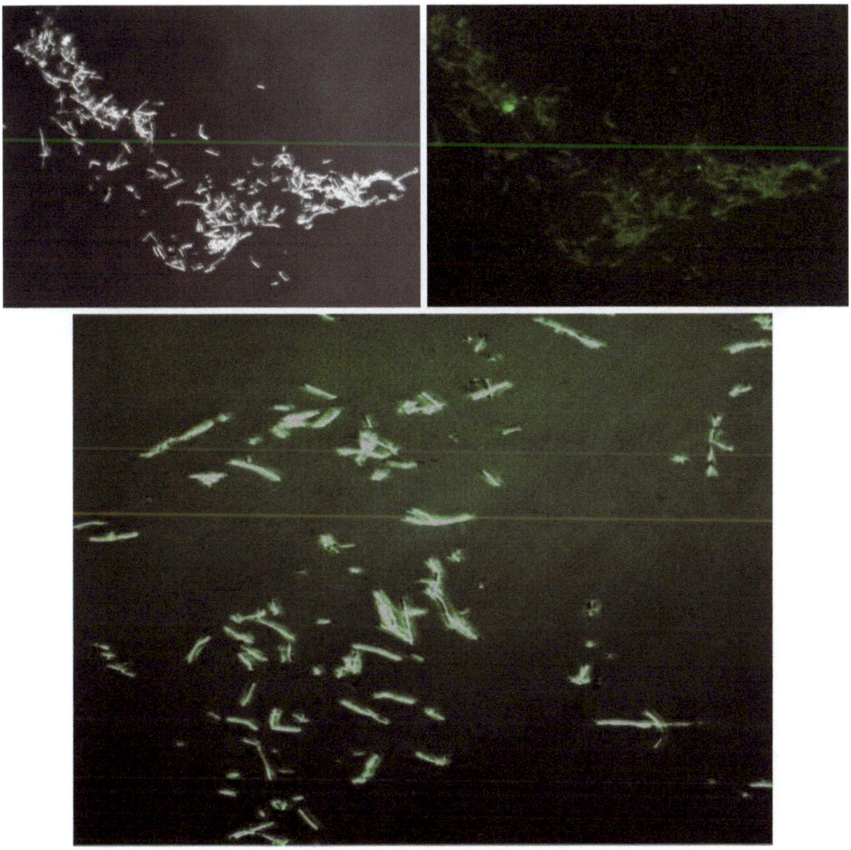

Fig. 3.24 Confocal fluorescence image of CNTs (d = 140 nm) treated with DCTP (*top*), and treated with Pyranine (*bottom*). The *darker background* is the silicon substrate

Using Pyrene-COOH Linker

Physical adsorption can even be strongly enhanced with the aid of aromatic molecules, for instance, pyrene linkers which develop strong π–π interactions with the graphitic sidewalls of carbon nanotube [22, 23]. This kind of functionalization has been very useful for carbon sensors in a Field-effect transistor configuration in which, as mentioned previously, the sp [2] network has to be preserved to assure the CNT conductance.

A conclusive and common way to confirm adsorption of pyrenes via π–π stacking is by fluorescence. The pyrene immobilization was followed by laser confocal microscope by using fluorescent pyrenes (dichlorotriazinepyrene, DCTP, fluorescence (λ_{exc} ~ 458 nm); pyranine, fluorescence (λ_{exc} ~ 460 nm)) as can be observed in Fig. 3.24.

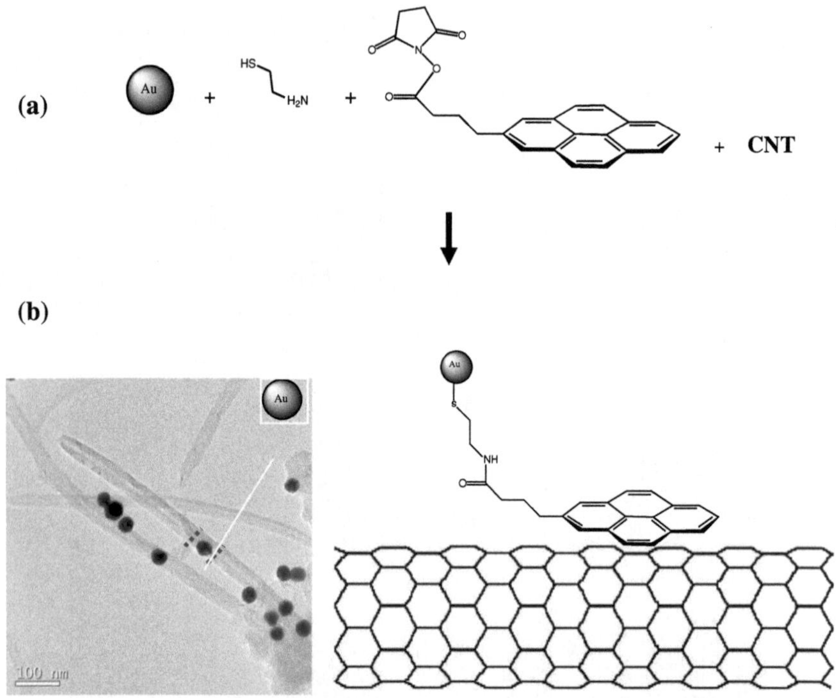

Fig. 3.25 **a** Formation of amide bonds: pyrene ester group reacts with the aminothiol cysteamine. **b** Scheme and TEM images of MWNTs functionalized. The *round dark spots* are the Au nanoparticles (~40 nm). CNT d = 30 nm

Another way to confirm adsorption of pyrenes is by using the functionalization scheme based in a bifunctional molecule, 1-(pyrene)butyric acid N-hydroxysuccinimide ester (1-PBASE), strongly adsorbed onto the sidewall of CNT via π–π stacking of the pyrene group. Its active ester group reacts with the amine groups, to form amide bonds; this can also be used to attach small molecules or biochemically active molecules (e.g. for protein immobilization). Figure 3.25 shows a TEM image of the system CNT-PBASE-cysteamine-Au nanoparticle that allows us to confirm the interaction of the 1-pyrenebutyric acid N-hydroxy-succinimide ester with carbon nanotube sidewall via π-stacking.

In order to evaluate non-specific adsorption processes, control experiments are necessary. TEM images of Fig. 3.26 show the results of different strategies of functionalization schemes which help to assess the extent of non-specific adsorption.

In images a and b high density of Au nanoparticles is observed. This is due to the presence 1-PBASE + cysteamine, following the functionalization scheme of Fig. 3.25. From image c it can be deduced that Au itself does not adsorb on CNTs as well as if there is negligible adsorption of cysteamine, d.

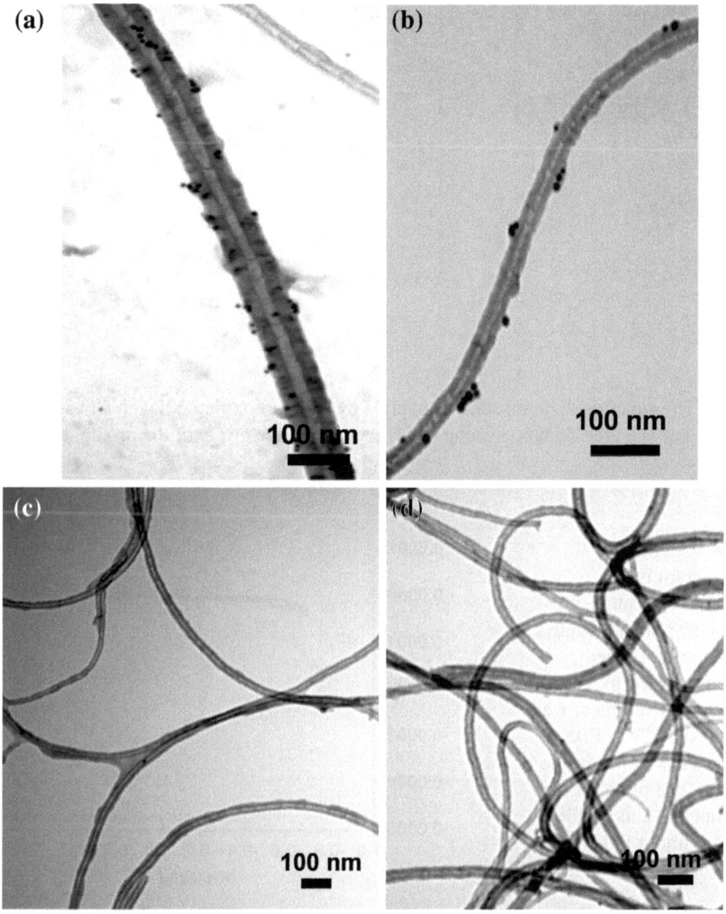

Fig. 3.26 **a** and **b** TEM images of bamboo-like MWNTs (diameter 30 nm) functionalized with gold nanoparticles (5 nm) through the 1-PBASE linker and cysteamine, **c** CNT-Au, **d** CNT-1-PBASE-Au without cysteamine

Another example of derivatization protocol was carried out by using a linker based on pyrene with a carboxylic functional group. The electrode, e.g. a multi-walled carbon nanotube composite was dipped on 6 mM solution of pyrenebutyric acid (at pH ~8) for 2 h and thoroughly rinsed in water. At this stage, the carboxylated pyrene is adsorbed on carbon nanotubes sidewall and can be used to covalently anchor other biomolecules.

In this case, the derivatization was tested by monitoring the electrochemical response of the system associated to the immobilization of biomolecules. For covalent attachment, the electrode was incubated for 24 h in a phosphate buffer solution (PBS, pH = 7) containing 6 mg/ml of the biomolecule, e.g. myoglobin

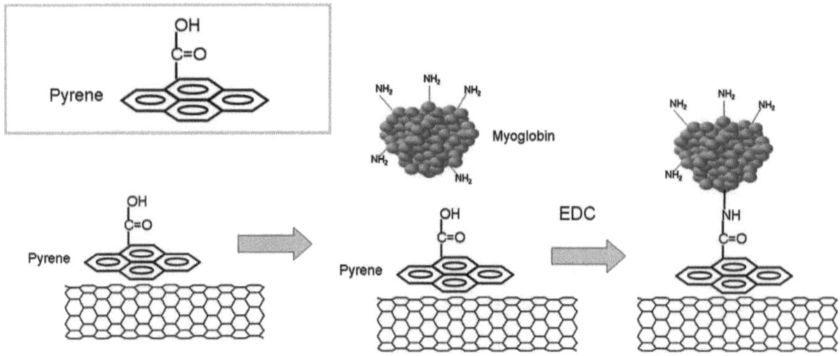

Fig. 3.27 The functionalization scheme using a pyrene carboxylated linker. Covalent attachment takes place between the carboxyl group of pyrene and amine group of the protein (myoglobin)

Fig. 3.28 Cyclic voltammogram for a hollow MWNT composite (diameter 30 nm and length 5–15 μm) immobilized with Mb through a pyrene linker. The measurement was taken at 100 mV/s in PBS solution (pH = 7) in aerobic conditions. The plots for the bare composite and for the CNT/Mb without linker are also shown

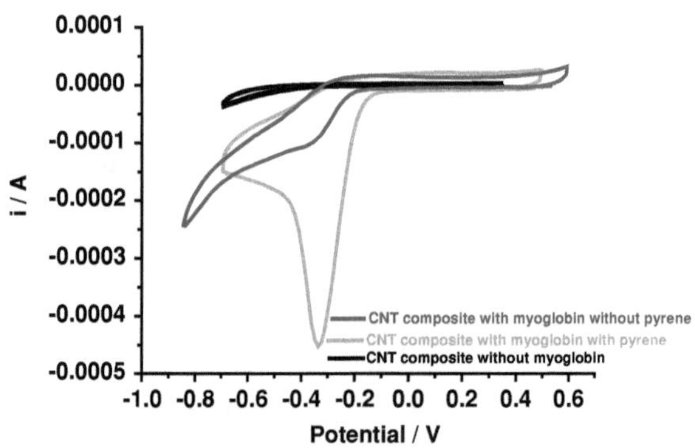

and 1 mg/ml of EDC. After that, the electrode was rinsed for one hour in PBS, and, is ready to be used. Figure 3.27 shows the derivatization protocol.

Figure 3.28 shows the voltammogram corresponding to the carbon nanotube composite modified with Mb via the pyrene linker. The i/V response of the composite with Mb but without linker and the bare CNT composite are also shown.

In the figure, it can be observed that the bare electrode shows no electrochemical response in the potential range, if the CNT electrode is functionalized with myoglobin without the pyrene linker, the electrochemical signal becomes remarkably weaker. However, the electrode modified with pyrene + Mb, exhibits a stronger reductive peak. This result confirms that the pyrene linker dramatically enhances the protein immobilization without compromising its redox activity.

However, proteins can also be non-specifically bound to the external sides of the carbon nanotube walls. This is why sidewalls of carbon nanotubes must be coated with polymers (e.g., with polyethyleneglycol, PEG) preventing the non-specific adsorption of proteins as will be mentioned later.

3.4 Materials and Equipment

3.4.1 Methods of Detection: Electrochemical and Electronic Biorecognition Processes

In few lines we will specify the materials and equipment used for the electrochemical and electronic detection of biorecognition processes as well as the buffer solutions used.

3.4.1.1 Electrochemical Detection

For the electrochemical characterization we used different electrochemical techniques: cyclic voltammetry, impedance spectroscopy, differential pulse voltammetry and chronocoulometry.

All measurements were performed with the normal three-electrode configuration. A Pt wire was used as counter electrode and an Ag/AgCl electrode was acting as reference electrode (in close proximity to the working electrode to minimize the ohmic drop). The working electrodes were bare/ADN/Protein/Aptamer modified carbon electrodes.

Cyclic voltammetry (CV) was mainly performed at a scan rate of 0.1 V s^{-1} generally in presence of redox indicators such as $[Fe(CN)_6]^{3-/4-}$, $Ru(NH_3)_6{}^{2+/3+}$ and at their appropriate potential range.

CV and impedance measurements were performed by using two different electrochemical workstations: IME6, Zahner, Germany and the potentiostat/galvanostat instrument form Metrohm Autolab. The impedance spectra were recorded in the frequency range 0.1 Hz–100 kHz at the redox equilibrium potential. The signal AC amplitude to perturb the system was 10 mV. Sampling rate was of 10 point per decade above 66 Hz, and 5 points per decade at lower range (at Zahner workstation). The obtained spectra were represented as Nyquist plots ($-Z_{imag}$ vs. Z_{real}) in the complex plane.

The experimental data were fitted to equivalent circuits which allowed obtaining the electrical parameters (resistances, capacitances, etc.) that can describe the electrochemical interface.

In all cases, the impedimetric spectrum of the bare electrode (blank) was recorded in first place. Then, impedance measurements were acquired after each further electrode surface modification (e.g. probe immobilization, hybridization with target, etc.).

The cyclic voltammetry and impedance spectroscopy were performed in electrolyte solutions that consisted mainly of 0.01 M potassium ferricyanide $K_3[Fe(CN)_6]$, 0.01 M potassium ferrocyanide $K_4[Fe(CN)_6]$ and 0.1 M KCl. When working with biological molecules, experiments were realized at physiological pH using PBS (pH ~ 7.4) buffer containing 0.02 M K3[Fe(CN)6]/K4[Fe(CN) 6] (1:1). For some experiments, 0.001 M $Ru(NH_3)_6Cl_3$ in 0.1 M KCl or 1.8 mM

$Ru(NH3)_6^{3+}$ in 10 mM TRIS + 10 mM NaCl buffer solution ((pH ~ 7.) were also used.

Differential pulse voltammetry (DPV) experiments were performed in 1.0 mM $K_4[Fe(CN)_6]$ and 1.0 M KCl at a pulse amplitude of 25 mV with an interval time of 1.2 s and a modulation time of 50 ms.

All chronocoulometry experiments were performed in Autolab electrochemical workstation in 10 mM TRIS + 10 mM NaCl or in the same electrolyte with 50 μM of $Ru(NH_3)_6^{3+}$.

Ruthenium hexamine, $Ru(NH_3)_6^{3+}$and potassium ferro/ferricyanide, $Fe(CN)_6^{3-/4-}$ were purchased from Sigma (MA, USA), and prepared in doubly distilled water. Other reagents (salts for preparation PBS, etc.) were commercially available and were all analytical reagent grade. The solutions were made using Milli-Q water.

3.4.1.2 Electronic Detection

Electrical characterization of FET devices and biorecognition process were performed by measuring the source drain current, I_{SD} as a function of the gate voltage (V_G) at a fixed source drain voltage (V_{SD}) of 100 mV. Typical values of V_G were ranging from −1 to 1 V in aqueous environment and from −5 to 5 V in dry environment. The setup consisted in a SUSS Microtec PA200 SemiAutomatic Probe and a Semiconductor Parameter Analyzer (4155B, Hewlett Packard). In this case, the sensing elements have to be prepared in low buffer concentration, 0.05× PBS, to provide an appropriate Debye length which can ensure the sensing process.

3.4.1.3 Buffer Solutions

The 1× PBS consisted of NaCl (50 mM), KCl (2.7 mM), Na_2HPO_4 (10 mM), KH_2PO_4 (1.76 mM). The pH was adjusted to 7.4 by adding HCl or NaOH. K^+ ions were necessary for the binding buffer solution because it stimulates the G-quartet assembly of thrombin aptamer [24].

TRIS, tris(hydroxymethyl)aminomethane, with the formula $(HOCH_2)_3CNH_2$ is another widely used buffer solution with pK_a of 8.06 which implies that the buffer has an effective pH range between 7.0 and 9.2 coinciding with the typical physiological pH of most living organisms. We have used it as an electrolyte for electrochemical measurements at a concentration of 10 mM.

3.4.2 Materials for Preparation of Carbon Working Electrodes

In this section, the materials used for preparation of Carbon-based composites, vertically aligned CNTs and other carbon systems will be listed.

3.4.2.1 Carbon-Based Composites

Different sources of carbon nanotubes were used.

The carbon nanotube composite electrodes were prepared with:

- Single-walled carbon nanotubes (SWNTs) prepared by chemical vapour deposition (CVD) with lengths between 5 and 20 μm, diameter about 1.3 nm and purity > 90 % were purchased from Helix Materials (Richardson, TX, USA).
- Multi-walled carbon nanotubes (MWNTs) prepared by CVD with lengths between 5 and 15 μm, diameter of 30 and 50 nm and purity 96–98 % were obtained from SES Research (Houston, TX, USA).
- MWNTs prepared by CVD with lengths between 5 and 15 μm, diameter about 140 nm and purity 90 % were obtained from MER Corporation (Tucson, AZ, USA).
- MWNTs in a bamboo structure prepared by CVD with lengths between 5 and 20 μm, diameter about 30 nm and purity 95 % were obtained from Nano-Lab (Newton, MA, USA).

The graphite-epoxy composites were prepared by using graphite powder of 10–20 μm particle size.

For the composite fabrication, Epotek H77A conductive resin and hardener Epotek H77B (both from Epoxy Technology, Billerica MA, USA) were used in all cases.

3.4.2.2 Vertically Aligned CNTs

The forest CNT macroelectrodes were provided by MER Corporation (Tucson, AZ, USA). Such electrodes were made of high density multi-walled carbon nanotubes (about 0.2 g cm^{-3}) grown on a silica substrate by chemical vapour deposition. The thickness of the resulting nanotube film was around 30–40 nm with a nanotube diameter of 30–40 nm.

For the development of CNT-ultramicroelectrode arrays, and carrying out the "opal inversion" procedure, we have used polystyrene latex particle of 900 and 600 nm diameter from Duke Scientific, CA and 800 nm from Kisker Biotech (10 % aqueous dispersion). The conductive substrates were doped (n-type) silicon wafers (100), which contain arsenic as a dopant (resistivity 0.002–0.05 ohm cm) from Cemat Silicon S.A. (Warsaw, Polland).

Other reagents such as Triton X-100, methanol, dichloroethane and dichloromethane were supplied by Sigma Aldrich.

Electrochemical pretreatments of carbon nanotube electrodes were performed using 1 M HNO_3 (Panreac).

3.4.2.3 Other Carbon Systems

For comparison, other electrodes surfaces were used such as Au coated silicon wafers and Highly Oriented Pyrolytic Graphite (HOPG, SPI supplies, PA, USA).

For the experiments with basal HOPG, highest HOPG quality (SPI Grade 1) was purchased from SPI supplies (West Chester, PA, USA). The edge HOPG was measured by taking the high quality HOPG and reorienting it perpendicularly in such a way that the edges of the layers of graphite were in contact with the electrolyte

3.4.3 Electrode Modification for Biorecognition Events

For functionalization we have used N-(3-dimethylaminopropyl)-N′-ethylcarbodiimide hydrochloride (EDC) from Sigma (St. Louis, MI); N-hydroxysulpho-succinimide sodium salt (Sulfo-NHS), from Fluka (Buchs, Switzerland) and hydroxylamine hydrochloride from Merck (Darmstadt, Germany).

For blocking the non-specific adsorption of the biomolecules, sodium dodecylsulphate (SDS) was purchased from Sigma (St. Louis, MI) and poly(ethylene glycol) (PEG) was purchased from Fluka (Buchs, Switzerland).

For passivation of FET device, polymethyl methacrylate (PMMA) with a molecular weight of 950000 and a copolymer of PMMA with a certain percentage of methacrylic acid (8.5 %) were from MicroChem Corp (Newton, MA).

Other reagents were commercially available and were all analytical reagent grades. The aqueous solutions were made using Milli-Q water.

3.4.3.1 Iron Protein Based Amperometric Biosensors

Horse heart myoglobin (Sigma, MA, USA) and Catalase from bovine liver (Sigma, MA, USA) were used as received and were dissolved in 0.1 M phosphate buffer solution (PBS, pH = 7.4).

For the study of the pH effects on the catalytic reduction of oxygen, phosphate buffer pH was adjusted with 2 M HCl or NaOH solutions. The pH range studies were from 5.0 to 9.40. Also, sodium tetraborate (0.1 M, pH = 9.18) and Acetic/Acetate buffer solution (0.1 M, pH = 5.0) were used for the study. All the solutions were prepared with doubly-distilled water.

3.4.3.2 DNA-Detection

The Different oligonucleotides were synthesized by TIB-MOLBIOL (Berlin, Germany). Their sequences and modifications were the following: probe-NH_2: 5′-NH_2-CCCCCCCCCCCCCCCCCCCC-3′ (Poly-C-NH_2-20 bases); complementary target: 5′-GGGGGGGGGGGGGGGGGGGG-3′ (Poly-G-20 bases): non-complementary target: 5′-TTTTTTTTTTTTTTTTTTTT-3′ (Poly-T-20 bases). Stock solutions of the oligonucleotides were diluted with sterilized and deionized water, separated in fractions and stored at −20° C. When required, a single fraction was defrosted and used

3.4.3.3 Aptasensor

The synthetic sequence of the DNA-lysozyme aptamer designed by Ellington et al. [25]. (5′- NH_2-ATC TAC GAA TTC ATC AGG GCT AAA GAG TGC AGA GTT ACT TAG- 3′) was amine 5′ modified and purchased from Invitrogen S.A. The specific protein, Lysozyme from chicken egg white and non-specific ones, bovine serum albumin (BSA), cytochrome C (CyC) and myoglobin (Mb) from equine heart were obtained from Sigma–Aldrich.

The thrombin aptamer-modified at the 5′ end with amine linker, 5′-NH_2-TTT-GGT TGG TGT GGT TGG- 3′ was provided by Invitrogen S.A. α-Thrombin from bovine plasma was provided by Sigma.

3.5 Additional Equipment

For the growth of carbon nanotubes and to characterize them, we have used additional equipment which will be listed next.

3.5.1 Growth of Carbon Nanotubes

For some specific studies we have performed CNT growth on silicon substrates by using a chemical vapour deposition furnace Lindberg blue with a tubular quartz reactor.

The CNT growth of carbon nanotube microelectrode arrays was performed in a JetStar 100ST Rapid Thermal CVD from Jipelec.

The plasma enhanced chemical vapour deposition of VACNTs was carried out in PECVD reactor which was developed by the FEMAN group (Universitat de Barcelona) as a result of an European project.

3.5.2 Characterization Techniques

We have used different characterization tools for the morphological and structural characterization of the different CNT-based systems. On one hand, we have worked with electron based techniques such as scanning electron microscopy (SEM) and Transmission Electron Microscopy (TEM) and on the other hand, scanning probe techniques based on Atomic Force Microscopy (AFM).

The nanotube morphology (shape, diameter and size) was observed by a scanning electron microscope (model s-570, HITACHI, Japan). In order to provide adequate resolution images, the sample were prepared properly. The easiest way was to place the sample on a metal support (place a drop of solid, CNT, removing the solvent in suspension by a stream of nitrogen gas)

Energy-dispersive X-ray spectrometer (EDS), located on the scanning electron microscope, allowed the verification of the metal nature of the nanostructures.

The structural characteristics of the nanotubes, the presence of amorphous carbon and catalyst particles were observed by a transmission electron microscope (Model JEM-2011, JEOL, Japan). Samples for TEM analysis were prepared by sonicating a small amount of nanotubes dissolved, for example, in two millilitres of 1, 2-dichloroethane or N, N-dimethylformamide. A drop of the resulting suspension was deposited on a copper grid with carbon coating and allowed to evaporate.

Topography and conductivity mappings were taken with an atomic force microscopy operated in intermittent mode and in current sensor mode, respectively (PicoSPM, Molecular Imaging). AFM allowed us the study of the morphology of the samples, determination of lengths and diameters, observation of metal particles, observation of the assembly of polystyrene spheres, etc.

In order to have more insights on the structural characteristics of these electrodes, we also performed Raman measurements. Micro-Raman spectra were collected using a wavelength excitation of 514.5 nm (T64000, Jobin Yvon).

A common way to confirm functionalization is by fluorescence. We have used a laser confocal microscope from Leica (TCS SP2 AOBS) Heidelberg, Germany. Such microscope was used to follow up pyrene functionalization.

We also used X-ray photoelectron spectroscopy (XPS) to measure chemical composition of our samples and to verify bond formation during the functionalization process. The system was a Kratos Axis ULTRA.

BET measurements, which provided information about specific surface area of the CNT platforms and porosimetry, were taken with a Micromeritics model Tristar 3000 equipment.

References

1. Cespedes F, Martinez-Fabregas E, Alegret S (1996) Trends Anal Chem 15:296–304
2. Bonanni A, Esplandiu M, Pividori M, Alegret S, del Valle M (2006) Anal Bioanal Chem 385:1195–1201
3. Pumera M, Merkoci A, Alegret S (2006) Sens Actuators B Chem 113:617–622
4. Wang S (2006) Nanotechnology 17:1551–1557
5. Martin-Fernandez I, Gabriel G, Rius G, Villa R, Perez-Murano F, Lora-Tamayo E, Godignon P (2009) Microelectron Eng 86:806–808
6. Martín-Fernández I (2010) Doctoral thesis, Universitat Autònoma de Barcelona, Bellaterra
7. Zhong G, Iwasaki T, Honda K, Furukawa Y, Ohdomari I, Kawarada H (2005) Jpn J Appl Phys Part 1 Regul Pap Short Notes Rev Pap 44:1558–1561
8. Xia Y, Gates B, Yin Y, Lu Y (2000) Adv Mater 12:693–713
9. Deckman HW, Dunsmuir JH (1982) Appl Phys Lett 41:377–379
10. Giersig M, Mulvaney P (1993) Langmuir 9:3408–3413
11. Hulteen JC, Van Duyne RP (1995) J Vac Sci Technol A Vac Surf Films 13:1553–1558
12. Kosiorek A, Kandulski W, Chudzinski P, Kempa K, Giersig M (2004) Nano Lett 4:1359–1363
13. Tu Y, Huang ZP, Wang DZ, Wen JG, Ren ZF (2002) Appl Phys Lett 80:4018–4020

14. Ren ZF, Huang ZP, Xu JW, Wang JH, Bush P, Siegal MP, Provencio PN (1998) Science 282:1105–1107
15. Terrones M (2003) Annu Rev Mater Res 33:419–501
16. Bower C, Zhu W, Jin S, Zhou O (2000) Plasma-induced alignment of carbon nanotubes. Appl Phys Lett 77:830–832
17. Kong J, Soh HT, Cassell AM, Quate CF, Dai H (1998) Nature 395:878–881
18. Katz E, Willner I (2004) Angewandte Chemie Int Ed 43:6042–6108.
19. Céspedes F, Alegret S (2000) Trends Anal Chem 19:276–285
20. Ambrosi A, Castañeda MT, Killard AJ, Smyth MR, Alegret S, Merkoçi A (2007) Anal Chem 79:5232–5240
21. Zacco E, Pividori MI, Llopis X, del Valle M, Alegret S (2004) J Immunol Methods 286:35–46
22. Besteman K, Lee J-O, Wiertz FGM, Heering HA, Dekker C (2003) Nano Lett 3:727–730.
23. Dai H (2002) Acc Chem Res 35:1035–1044.
24. Spiridonova V, Rassokhin T, Golovin A, Petrova E, Rohzdestvensky T, Pakhomova Y, Kopylov A (2002) Bioelectrochemistry 56:95–97
25. Kirby R, Cho EJ, Gehrke B, Bayer T, Park YS, Neikirk DP, McDevitt JT, Ellington AD (2004) Anal Chem 76:4066–4075.

Chapter 4
Results and Discussion: Impact of Nanotechnology in Sensors

A sensor is a device that responds to a stimulus by generating a functional output induced by a change in some intrinsic properties. We are surrounded by sensors and sensing networks that monitor a multitude of parameters in view of enhancing our safety and quality of life. Sensors assist us in health care and diagnostics, they monitor our environment, our aeroplanes and automobiles, our mobile phones, game consoles and watches, and last but not least, many of our human body functions. Modern sensing systems have greatly benefited in recent decades from advances in microelectronics and microengineering, mainly in view of making sensors smaller, cheaper, more sensitive, more selective, and with a better signal-to-noise ratio, following classical scaling rules.

Nanoscale features have a great impact on many sensing systems, in particular where the surface-to-volume ratio plays a fundamental role, such as in certain chemical and gas sensors. The high surface-to-volume ratios of nanoporous and nanostructured materials have led to their implementation in sensing systems since sensing research first began to engage with the nanotechnology. The surface plasmon resonances of nanostructures have also enriched the scope for developing novel sensing devices. On the other hand, sensors where bulk properties dominate, such as inertial sensors, are less likely to benefit from extreme scaling.

Advances in thin film techniques and chemical synthesis have allowed material properties to be tailored to sensing requirements for enhanced performance. These bottom-up fabrication techniques enable parallel fabrication of ordered nanostructures, often in domain-like areas with molecular precision. At the same time the progress in top-down methods such as scanning probe lithography, nanoimprint lithography, soft-lithography and stencil lithography have also facilitated research into sensing and actuating nanotechnology. While being radically different from each other, these techniques represent a formidable toolset for structuring materials at the nanoscale in a multitude of fashions. The availability of these new nanopatterning techniques are increasingly implemented in the manufacturing of advanced sensor systems, and we can expect in the next decade an increased emergence of micro- and nanosensor systems that

M. Pacios Pujadó, *Carbon Nanotubes as Platforms for Biosensors with Electrochemical and Electronic Transduction*, Springer Theses, DOI: 10.1007/978-3-642-31421-6_4,

implement novel nano-functionalities thanks to cost-effective fabrication. More and more researchers, engineers, and even schoolchildren, can benefit from and use these new methods and devise novel applications for nanosystems. This is certainly beneficial to expediting a further dramatic increase in knowledge and the development of actual devices and applications that put gains in our understanding of nanosystems into practice. Nanotechnology is a relatively young discipline compared to classical engineering, and it is inherently interdisciplinary. It seems that in many fields we are actually just beginning to venture into these new dimensions. Challenges remain, however, in all aspects of nanotechnology. We need to improve imaging performance by enabling faster (video rate) coverage of larger surfaces, eventually down to the molecular scale. We also need to perfect nanopatterning methods to improve resolution, overlay and throughput capabilities. Future nanomanufacturing will most likely rely on combinations of top-down engineering and bottom-up self-assembly. Last but not least, we need to find ways for the mutual integration of multiple length-scale devices (nano/micro/macro) so that we can program a 'nano-functionality' into a microsystem exactly where it is needed. Such improvements will ultimately lead to improved sensors and contribute not only to improvements in our quality of life but also to building energy-saving systems that can be fabricated with low-waste manufacturing methods.

This article adapted from Brugger (2009) [1] summarizes the motivation and progression of this thesis, towards the implementation of nanostructured materials (carbon nanotubes) in sensing devices. As starting point, we have incorporated these nanomaterials to electrode platforms of similar format to the one that has been widely used in our lab, the macrosystem Graphite-Epoxy Composite (GEC). Graphite-epoxy composites have been successfully used in the development of genosensors [2], immunosensor [3] or amperometric sensors incorporating amino acids or enzymes [4]. They have been reported to have enhanced electrochemical properties compared with other carbon electrodes (as glassy carbon electrodes) attributed to the random assembly of graphite particles working as an array of microelectrodes and to the non-conductive rigid epoxy regions of the electrode [5].

However, the exceptional properties of carbon nanotubes open new possibilities to improve the overall properties of conventional materials for the development of biosensors. All the solved and unsolved challenges will be described in this *Results and discussion* chapter which has been divided in three blocks.

We will focus mainly on two different aspects of the carbon nanotube electrochemistry. First, we will discuss the use of carbon nanotubes as electrochemical transducers and the relation between their structure and their electrochemical reactivity. The second issue deals with the use of carbon nanotubes for (bio)sensing and the use of electrochemistry as a signature of the carbon nanotube functionalization and for following up a biorecognition event. Finally, we will highlight the use of carbon nanotube in field-effect configuration also for biosensor purposes.

References

1. Brugger J (2009) Nanotechnology 20:430206
2. Pividori MI, Merkoci A, Alegret S (2003) Biosens Bioelectron 19:473–484
3. Santandreu M, Cespedes F, Alegret S, Martinez-Fabregas E (1997) Anal Chem 69:2080–2085
4. Serradell M, Izquierdo S, Moreno L, Merkoçi A, Alegret S (2002) Electroanalysis 14:1281–1287
5. Ramirez-Garcia S, Alegret S, Cespedes F, Forster RJ (2004) Anal Chem 76:503–512

Chapter 5
Results and Discussion: Response of Different Carbon Platforms as Electrochemical Transducers

One of the most important aspects for evaluating the characteristics of an electrochemical transducer is based on the analysis of electron transfer kinetics. Generally, this phenomenon depends on the electrode materials and surface properties. Carbon-based materials such as glassy carbon, graphite, fullerene and doped-diamond, with different electronic and structural properties, have been shown to have different electrochemical properties [1]. A relationship between structure, morphology and electrochemical reactivity has been established and therefore it can be predicted that different structural forms of the CNTs will exhibit different electrochemical behaviour compared with other forms of carbon-based materials.

Indeed, an important number of publications have recognized the differences in the electrochemical reactivity between the CNT modified electrodes and control electrodes which is attributed to the high specific surface area of the CNTs and to the differences in their surface structure chemistry [2–7].

Therefore, the first part of this work has been focused on the study of the transduction capabilities of different systems based on CNTs to contribute with more insights in such particular issue. Specifically, the electrochemical performance of different rigid carbon nanotube/epoxy composites will be characterized.

All the composites were always prepared with the same weight ratio but different nature of carbon nanotubes: bamboo-like, single and multi-walled carbon nanotubes of different diameters (30, 50 and 140 nm). These systems were compared with other forms of carbon such as analogue graphite composites and HOPG (basal and edge). The analysis was focused on the electron transfer rate constants and charge transfer resistances; these parameters are related to the electrode responsiveness and sensitivity, two of the requirements to be considered in sensor design.

For that electrochemical evaluation, we have used cyclic voltammetry in presence of benchmark redox systems such as $Fe(CN)_6^{3-/4-}$. Such redox couple was chosen due to its wide use in the electrochemical community, which facilitates the comparison with different systems reported in literature.

M. Pacios Pujadó, *Carbon Nanotubes as Platforms for Biosensors with Electrochemical and Electronic Transduction*, Springer Theses, DOI: 10.1007/978-3-642-31421-6_5,

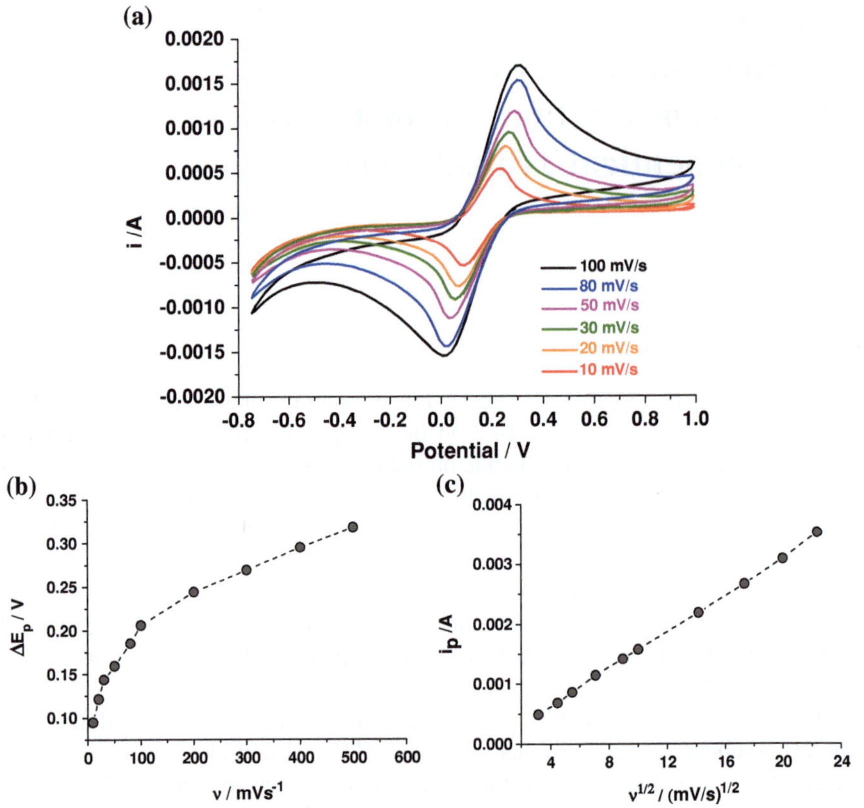

Fig. 5.1 Redox peak separation. **a** Cyclic voltammograms in $Fe(CN)_6^{3-/4-}$ at different scan rates; **b**, **c** Anodic peak current and redox peak separation as a function of the scan rate for a SWNT electrode as representative of the CNT composite behaviour

The two main features to analyse in a cyclic voltammogram are the peak current and the peak separation. The first one can give us insights on the macroscopic electroactive surface and the second one provides a measure of the heterogeneous reaction rate constant.

The electrochemical response of a SWNT composite as a function of the scan rate is illustrated in Fig. 5.1. The SWNT composite is taken as a representative system but it is worth mentioning that the same qualitative characteristics were also observed with the other CNT composites.

It can be observed that the peak separation is bigger than 59 mV; hence the system is not totally reversible. In the investigated scan range, the I_p varies linearly with the square root of the scan rate, $v^{1/2}$, and the peak separation becomes larger with increasing scan rate. These results reflect an electrochemical behaviour which is controlled by the electron transfer process.

In order to gain more insight on the relative electron transfer process itself, the electrochemical behaviour of CNT composites with the same geometric area

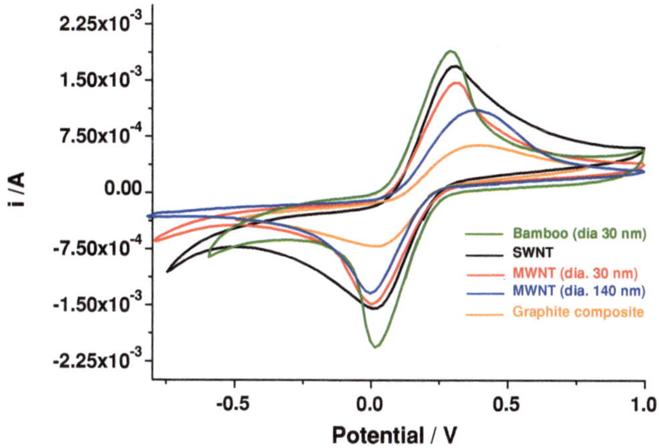

Fig. 5.2 Cyclic voltammograms of carbon composites. The i/V profiles were taken at 100 mV/s in presence of 0.01 M $Fe(CN)_6^{3-/4-}$ + 0.1 M KCl

(electrode area of 0.28 cm^2) was compared. i/V profiles were taken at the same experimental conditions for the different carbon nanotube composites and compared with the graphite composites (Fig. 5.2). Two features reveal the different electrochemical behaviour of these systems: the peak separation and the peak current. The graphite-epoxy composite exhibits the highest peak separation whereas in the carbon nanotube composites, the peak difference decreases with the diameter of the nanotube. A particular behaviour is found with the bamboo composite which with a diameter of 30 nm shows the smallest peak difference. One should recall that the peak separation gives us an estimation of the facility at which the electron transfer process is taken place and therefore, it can be used to extract information of the heterogeneous electron transfer rate constant: the greater the separation between peak potentials (ΔEp) of voltammogram the lower the electron transfer rate constant. Additionally, the peak current is remarkable enhanced in the case of the carbon nanotube composites as compared with the graphite ones. In turn, the peak current of the nanotubes composites increases as the diameter decreases, and the CNTs composite with the highest peak current is the one with bamboo-like nanotubes. The increase of peak currents can be ascribed to an increase of the electrochemical active surface area [8, 9].

Continuing with the carbon-composites comparison, topographical AFM and AFM conductivity mapping[1] characterization were performed.

Atomic force microscopy was used in order to get insight on the surface characteristics of the CNT composites in comparison also with the graphite-epoxy ones, Fig. 5.3. First of all it can be observed that the surface area of the CNT composites is larger than in the case of the graphite system. Indeed, the surface

[1] Conductivity mapping has been acquired by using AFM in the current sensor mode.

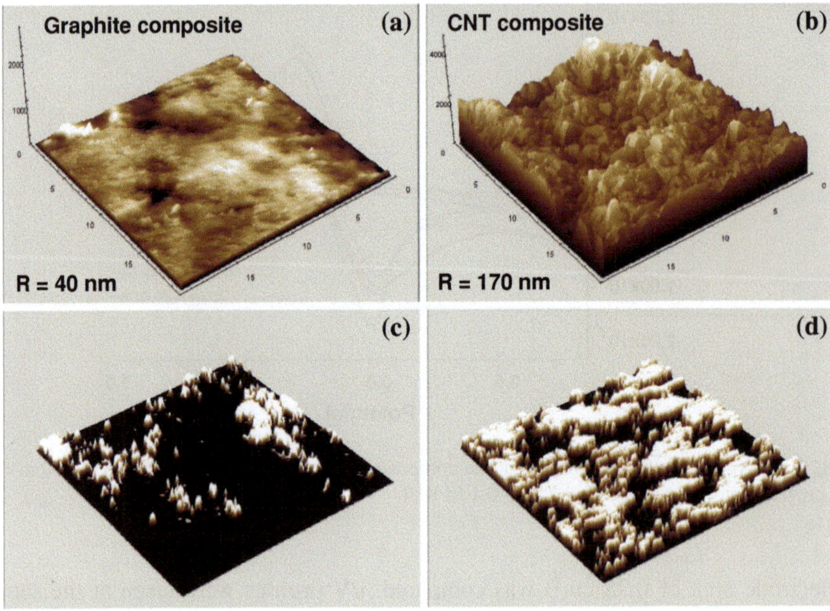

Fig. 5.3 Topographical AFM images of the graphite (**a**) and CNT composites (**b**). The CNT composite was prepared from the hollow multi-walled CNT. The roughness *R* for each system is also reported. On the *lower row*, the conductance mappings for graphite-epoxy (**c**) and for the CNT composite (**d**) are also depicted

roughness of the CNT composites is about four times the roughness of the graphite ones. By using the AFM in the current sensing mode, conductivity mappings of the composite surfaces were also performed (Fig. 5.3c, d). It is worth mentioning that both AFM images, the topographical and conductivity map were taken simultaneously. The conductivity data show no current response (dark regions) on the insulating epoxy regions but high current response on the carbon ones (lighter ones). From the comparison of the current images for both carbon composites, the one based on CNT shows more conductive areas than in the case of the graphite-epoxy. Therefore, the surface characterization performed with AFM seems to corroborate the expected increase in electrochemical active surface area for the case of the CNT composites.

The electrochemical surface area has been quantified from the CV measurements by calculating the peak current according to the relationship, $I_p = 3.01 \times 10^5\ n^{3/2}\ (\alpha D_{red} \nu)^{1/2} A C^{*}_{red}$, adequate for electron transfer-controlled processes. In this equation, α represents the symmetry factor which was taken approximately as 0.5, $D_{red} = 6.32 \times 10^{-6}\ cm^2/s$ corresponds to the diffusion coefficient of the reduced species, $\nu = 0.1$ V/s represents the scan rate, A the electrochemical active area and $C^{*}_{red} = 0.01$ M is the bulk concentration of the electroactive species. The calculated electrochemical surface areas are reported in Table 5.1.

Table 5.1 Peak separation, ΔE_p (corrected by ohmic drops) and electron transfer rate constants, k^o of different carbon composite systems calculated from cyclic voltammetry data by using Nicholson method

System	I_p(A)	A(cm^2)	ΔE_p (V)	k° (cm/s)^V	k° (cm/s)^I
CNT bamboo	2.00×10^{-3}	1.19	0.12	3.4×10^{-3}	3.0×10^{-3}
SWNT	1.60×10^{-3}	0.95	0.13	3.0×10^{-3}	2.5×10^{-3}
MWNT30	1.47×10^{-3}	0.87	0.16	1.8×10^{-3}	1.9×10^{-3}
MWNT50	1.34×10^{-3}	0.80	0.18	1.4×10^{-3}	1.5×10^{-3}
MWNT140	1.27×10^{-3}	0.76	0.29	4.7×10^{-4}	5.8×10^{-4}
Graphite	7.20×10^{-4}	0.43	0.31	3.5×10^{-4}	2.6×10^{-4}
HOPG basal	3.36×10^{-4}	0.20	0.59	2.5×10^{-5}	1.0×10^{-4}
HOPG edge	1.00×10^{-3}	0.60	0.09	7.0×10^{-3}	8.0×10^{-3}

k^o (cm/s)^V and k^o (cm/s)^I represent the heterogeneous rate constants extracted from voltammetry and impedance spectroscopy respectively

In order to have a more quantitative insight of the electron transfer process at the CNT composite/electrolyte interface, we have evaluated the electron transfer rate constant from the peak separation by using the Nicholson method [10]. The resulting reaction rate constants are also listed in Table 5.1. In the reported values the ohmic drop has been compensated by using the values of the resistance obtained by impedance spectroscopy technique, as discussed below. The table also includes the calculated values from experimental i/V profiles for basal and edge HOPG in order to get reference values for the analysis of the composite performance.

Nicholson method consists in solving a φ parameter that is related to the separation of peaks. This parameter is calculated numerically from the diffusion equations of electroactive species and the equation describing the electron transfer at the electrode (Butler–Volmer) and appropriate boundary conditions. From the parameter, one can get the value of k^o by the equation:

$$\varphi = \left[\left(\frac{D_0}{D_r}\right)^{1/2}\right]^{\alpha} k^o \Big/ \sqrt{D_0 \pi n F v / RT}$$

where $D_o = 7.63 \times 10^{-6}$ cm^2/s and $D_r = 6.32 \times 10^{-6}$ cm^2/s are the mentioned diffusion coefficients of oxidized and reduced species, in our case, the ferro/ferricyanide redox couple, α is again the symmetry factor of the reaction (in this case we have also used the value of 0.5), n the number of electrons, F Faraday's constant, v is the scan rate, R the ideal gas constant and T temperature.

The trend observed for the rate constants resembles very well the results obtained from the voltammograms of the carbon nanotube-based systems: the reaction rate constants are higher for the nanotubes with smaller diameter and for the bamboo-like CNTs. Indeed, the electron transfer rate for the bamboo system is one order of magnitude higher than the one for graphite composite. Moreover, the values of the rate constants for the carbon nanotube and graphite composites fall

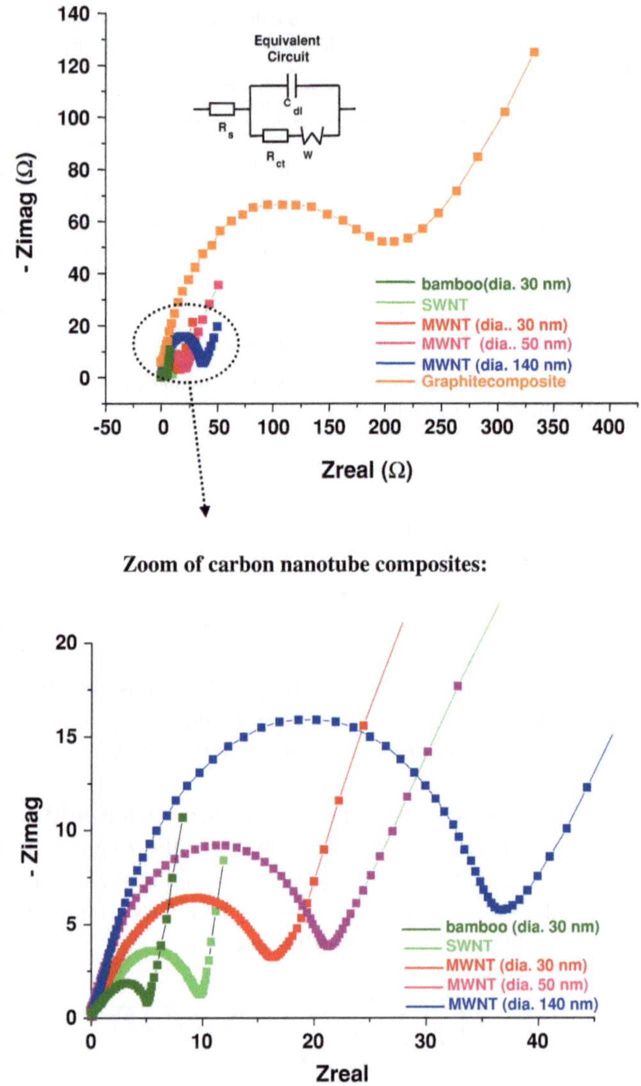

Fig. 5.4 Nyquist plot of different graphite and carbon nanotube composites in presence of the benchmark redox pair $Fe(CN)_6^{3-/4-}$. Zoom of Nyquist plot of carbon nanotube composites

in between the extreme values of edge and basal plane HOPG. The basal HOPG exhibits the lowest rate constant in agreement with previous reported values and at the same time presents more dispersion due to difficulty in controlling the edge density.

Impedance spectroscopy can also be used to complement the electrochemical reactivity of these electrode surfaces as shown in Fig. 5.4. Experimental impedance spectra in presence of the redox pair can be fitted according to the Randles equivalent circuit (inset Fig. 5.4) and extract information of the charge transfer resistance, R_{ct} (that corresponds approximately to the diameter of the semicircle).

Again, the same trend was observed as in the case of the cyclic voltammetry. Among these carbon-based composites, bamboo-like systems provides the lowest charge transfer resistance and the graphite composites, the highest one. In turn, the R_{ct} of the nanotube composites decreases with the diameter. For the same geometric area, the resistances of the bamboo CNT composites are around 6 Ω, SWNT around 10 Ω, for concentric MWCNT ones ranging from about 16 to 40 Ω and for the graphite-epoxy the resistances are higher than 200 Ω. These values are between the values obtained for basal HOPG (up to 800 Ω) and edge HOPG (below 5 Ω) as shown in Fig. 5.5.

From the impedance analysis, one can also extract the trend in the heterogeneous rate constant by using the following relationships, $R_{ct} = RT/nFi_0$ and $i_0 = FA\, k^o\, C^*$ [11], were i_0 corresponds to the exchange current, A is the electrochemical surface area calculated from the voltammetry, C^* is the bulk electroactive species concentration and R, T, F have their usual meaning. The heterogeneous rate constants calculated from impedance spectroscopy have been reported in Table 5.1. They exhibited relative errors below 20 %. From these values, one can observe a relatively good agreement with the ones extracted from cyclic voltammetry measurements. A higher dispersion was found for the basal HOPG, whose electrochemical properties are very difficult to measure, as it is difficult to control the density of edges.

All the findings obtained by cyclic voltammetry and impedance spectroscopy reveal the difference in electrochemical properties of carbon nanotube composites of different nature and characteristics.

The results suggest that as the density of the graphitic plane edges is increased, the electrochemical response gets enhanced. Since the composites were fabricated by using the same weight percentage, composites with smaller nanotube diameter will contain more carbon tubes than thicker diameter carbon nanotube composites.

Therefore, the percentage of nanotube edges per unit of electrode area will increase with decreasing carbon nanotube diameter and thus the electroactive sites for electron transfer processes.

Exceptionally, the 30 nm diameter bamboo-shaped carbon nanotube composites exhibit the highest electroactivity due to the high density of edge planes of graphene, the most reactive spots, along the wall.

These results seem to agree with the tight relationship between surface structure and electrochemical reactivity. The results also corroborate the previous work by Compton's group in which the electrochemical anisotropy of the CNTs and the role of edge planes as reactive sites for electron transfer were emphasized [12]. The open terminals of the nanotubes (similar to the edge planes of the graphite) present a fast kinetics for the electron transfer while the walls of the nanotubes (similar to the basal plane of the graphite) exhibit a slow kinetics for the electron transfer.

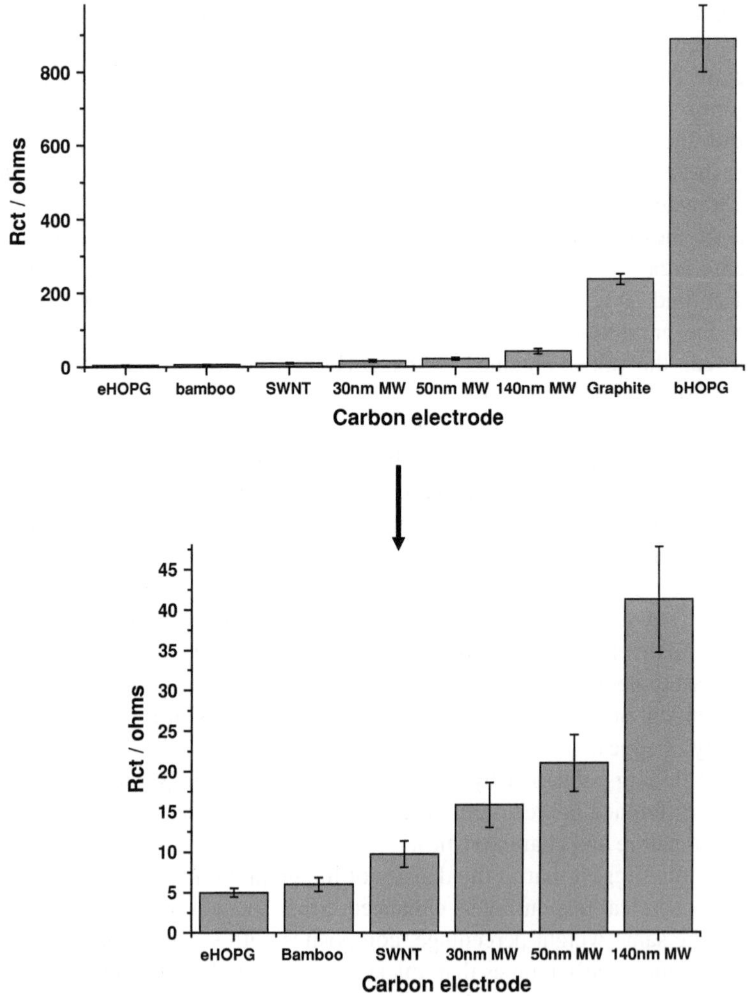

Fig. 5.5 Values of charge transfer resistance of different systems based on carbon, using the redox probe $Fe(CN)_6^{3-/4-}$. The *arrow* shows a zoom of the carbon nanotube-composite system and edge-HOPG

5.1 General Conclusions of Response of Different Carbon Platforms as Electrochemical Transducers

In this part, we have highlighted the use of carbon nanotubes-based materials as electrochemical transducers. The response of different carbon nanotubes composites has been analysed by focusing mainly on the kinetics of the electron transfer process and the charge transfer resistance. Therefore, our findings indicate that a high degree of CNT edges play an important role for increasing the electrochemical performance of CNT electrodes. Therefore, such issue has to be taken into consideration when designing sensitive CNT electrode platforms.

References

1. Gong K, Yan Y, Zhang M, Su L, Mao SXL (2005) Anal Sci 21:1383–1393
2. Moore RR, Banks CE, Compton RG (2004) Anal Chem 76:2677–2682
3. Banks CE, Davies TJ, Wildgoose GG, Compton RG (2005) Chem Commun 7:829–841
4. Banks CE, Compton RG (2006) Analyst 131:15–21
5. Musameh M, Wang J, Merkoci A, Lin Y (2002) Electrochem Commun 4:743–746
6. Wang J, Li M, Shi Z, Li N, Gu Z (2002) Anal Chem 74:1993–1997
7. Randhir PDJW, Nathan S (2005) Lawrence. Electroanalysis 17:65–72
8. Pumera M, Merkoci A, Alegret S (2006) Sens Actuators B Chem 113:617–622
9. Brito R, Tremont R, Cabrera CR (2004) J Electroanal Chem 574:15–22
10. Nicholson RS (1965) Anal Chem 37:1351–1355
11. Bard AJ, Faulkner LR (eds) (2001) Electrochemical methods: fundamentals and applications, 2nd edn. Wiley, New York
12. Banks CE, Davies TJ, Wildgoose GG, Compton RG (2005) Chem Commun 7:829–841

Chapter 6
Results and Discussion: Biorecognition Processes on Different CNT Platforms

The structural requirements of the carbon nanotubes for a better electrochemical performance have been analysed by considering CNT composites. Results have shown that a disposition with high density of edges is important for fast electron transfer. Therefore, the proper construction and orientation of the CNT electrodes will become very important. Consequently, a shift towards different concepts of electrode platforms can be an interesting strategy to follow. Promising alternatives can be the design and development of high density of vertically aligned carbon nanotubes (VACNT) or CNT-forest electrodes, vertically aligned carbon nanotubes microelectrodes (CNT microelectrodes) and ultramicroelectrodes (UMAs). Therefore, in this section we will start evaluating the electrochemical performance of different upright oriented CNT electrodes but also as supports for functionalization and biosensing. Specifically, different iron protein, DNA and aptamer-based sensors will be evaluated. All these studies have been pursued with the final goal of achieving an improvement of sensitivity and selectivity in the biorecognition process.

Finally, in the last *Results and discussion* section, we will also address another CNT configuration electrode, although it can be considered as an electrochemical sensor, its operating mode differs greatly from the ones conceived in this thesis. Specifically, we will evaluate CNT-FET configuration to follow biorecognition events in a very sensitive way.

6.1 Iron Protein-Based Amperometric Biosensors (Myoglobin and Catalase Response)

In this section, we will start evaluating forest or vertically aligned CNT electrodes and contrasting their response with CNT/epoxy, graphite/epoxy and other more conventional electrodes. For accomplishing that, we will use redox protein or enzyme modified electrodes motivated by the fact that enzymatic or redox protein

M. Pacios Pujadó, *Carbon Nanotubes as Platforms for Biosensors with Electrochemical and Electronic Transduction*, Springer Theses, DOI: 10.1007/978-3-642-31421-6_6,

biosensors represent a key research field in the wide area of electrochemical biosensors. Redox proteins or enzymes can not only been exploited for applications in the area of biomedicine, food technology or environment but can also become model systems for fundamental studies of remarkable importance in nature machinery. A better understanding of these basic processes can also help to mimic nature and to improve and exploit more efficiently sensor technology or develop bioreactors or novel fuel cells.

We have mentioned in Chap. 1 that electrochemical enzymatic sensors are based on the ability of a redox enzyme to transduce the turnover with its corresponding substrate into a detectable, reliable and quantifiable current. Therefore, different types of carbon-based systems have been used to anchor two metalloproteins, myoglobin (Mb) and catalase (Cat) and their catalytic activity has been evaluated.

6.1.1 Direct Electrochemistry

We will first reveal how CNT electrodes facilitate the direct electron transfer with myoglobin or catalase by using cyclic voltammetry in deaerated solutions of PBS. As an illustrative example Fig. 6.1 shows the cyclic voltammograms for the case of myoglobin covalently immobilized on three different carbon-based systems: graphite-epoxy composite, carbon nanotube epoxy composite, forest-like CNTs. These electrodes exhibit the typical anodic and cathodic peaks corresponding to the direct electron transfer between the Fe(III)/Fe(II) from the hemin core and the carbon-based electrodes. The formal redox potential was found to be −0.33, −0.29 and −0.37 V for the graphite-epoxy composite, forest-like CNTs and CNT epoxy composite, respectively. These values are in the range of those already reported for myoglobin on different CNT systems or on carbon electrodes modified with membranes [1–10]. The high background current for the CNT epoxy electrode, which could be mainly attributed to a higher roughness, is remarkable. However, the determination of precise roughness factors is rather troublesome, especially in the case of composite electrodes.

Nevertheless, some general trends concerning the relative roughness and porosity degree of the different electrodes can be obtained by calculating the capacitance of the double layer (C_{dl}). This parameter is sensitive to the effective electrochemically active area of the electrode which is proportional to the geometrical area with a proportional factor *r* (a factor that accounts for the roughness, porosity and heterogeneity of the electrode), i.e. $A = rA_{geom}$. We have extracted such a parameter by evaluating cyclic voltammograms as a function of the scan rate in the absence of faradaic reactions. The results of the C_{dl} weighted with the factor *r* for the different electrodes are listed in Table 6.1. One can observe that such values are noticeably high as compared to those of flat electrochemical interfaces (typically, a few $\mu F\ cm^{-2}$) [11]. This is an indication of the rough nature of the surface of the electrodes, which is especially important in the CNT composite. These results are in agreement with BET measurements shown in the experimental part.

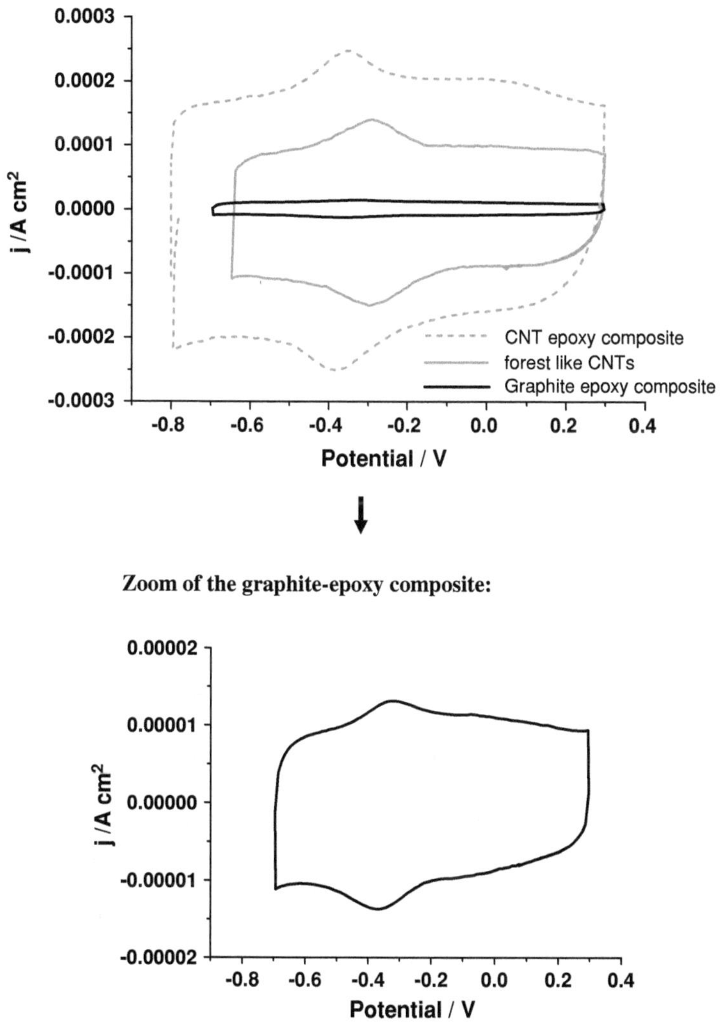

Fig. 6.1 Cyclic voltammograms for a forest-like CNT electrode, carbon nanotube epoxy composite (CNT, 30 nm diameter) and graphite-epoxy composite modified with Mb in deareated PBS solutions at a scan rate of 0.1 V/s. A zoom of the graphite composite response is also depicted

Table 6.1 Capacitance of double layer (C_{dl}) of the carbon-based electrodes

Carbon electrode	$C_{dl}r$ (F cm^{-2})
Graphite-epoxy composite	1.3×10^{-4}
Carbon nanotube epoxy composite	2.0×10^{-3}
Forest-like CNTs	9.7×10^{-4}

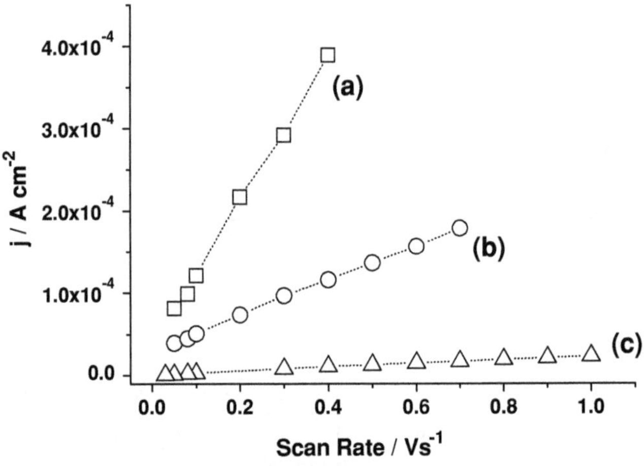

Fig. 6.2 Dependence of the oxidation peak current with the scan rate for the different carbon electrodes: **a** CNT composite, **b** forest-like electrode and **c** graphite composite

The important feature of the voltammetric profiles shown in Fig. 6.1 is that they exhibit a nearly reversible electrochemical behaviour which is reflected in the very small peak potential differences (ΔE_p) between the anodic and cathodic wave, a feature that is much remarkable in the case of Mb/forest-like CNT electrodes. The smaller the peak separation, the more facile is the electron transfer process. Additionally, the peak current is directly proportional to the scan rate as depicted in Fig. 6.2. All these characteristics are in agreement with the electrochemical behaviour of surface-confined electroactive monolayers. That means that a layer of myoglobin has been successfully immobilized at the surface. Under such conditions we can use Laviron Method for extracting electron transfer rate constants (k^o) which is applicable for diffusionless systems and based on the Butler–Volmer approach [12]. By numerically integration of the Butler–Volmer equations without considered diffusion, a Laviron plot of ΔE_p as a function of $n\nu F/RTk$[1] can be obtained for the special conditions of ΔE_p smaller than 200 mV and by assuming a value of the transfer coefficient, α, of 0.5. Figure 6.3 shows the Laviron plot under such conditions. We have used the value of ΔE_p at 0.1 V in order to extrapolate the electron transfer rate constant.

The ΔE_p together with the rate constants (k^o) values are reported in Table 6.2. We have found that forest-like CNTs exhibit the smallest peak separation and highest electron transfer rate constant followed by the CNT and graphite-epoxy composites. As mentioned, edge defects promote high electron transfer rates at the carbon electrode interface and it is expected that forest electrodes concentrate

[1] In this expression n is the number of electrons, ν is the scan rate and the rest of parameters have the usual meaning.

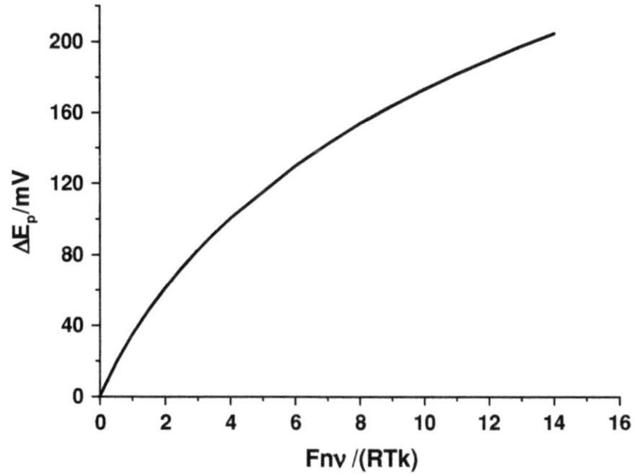

Fig. 6.3 Laviron plot for $\Delta E_p < 200$ mV and a transfer coefficient value of 0.5

Table 6.2 Electrochemical parameters of the carbon-based electrodes: peak potential differences (ΔEp), rate constants (k°), protein coverage per unit of geometrical area (Γr)

Carbon electrode	ΔE_p (mV)	k° (s^{-1})	Γr (mol cm^{-2})
Graphite-epoxy composite	43 ± 3	3.1 ± 0.3	6.8×10^{-11}
Carbon nanotube epoxy composite	21 ± 4	7 ± 1	8.0×10^{-10}
Forest-like CNTs	8 ± 3	15 ± 4	3.0×10^{-10}

a higher density of edges just at the interface facing the electrolyte which could favour a better protein wiring at the interface and promote the electron exchange. In the case of the CNT composites, the relatively lower electron transfer rate can be rooted on the more heterogeneous complexity of the porous CNT composite. Protein can be anchored on CNT composites through exposed edges on the CNT tips, on defects at the uncoated walls or on defects at insulated polymer coated CNTs. All these characteristics could explain the lower transfer rate as compared with the forest-like systems. Although the CNT epoxy composites have lower electron transfer kinetics, they exhibit a higher current signal.

That could be attributed to a higher material porosity providing more surface area for protein immobilization which is also in agreement with BET measurements.

An estimate of the protein surface coverage (Γ) for every electrode can be made from the peak current (i_p) as a function of the scan rate by using the relation corresponding to a reversible electroactive adsorbate layer, $i_p = 9.39 \times 10^5 n^2 \nu A \Gamma$. In this relation, n represents the number of exchanged electrons, ν is the scan rate, and A is the electroactive surface area which depends on the roughness of the surface ($A = rA_{geom}$). Since it is very hard to evaluate the true surface area and the coverage independently, we decided to report the protein coverage per unit of geometrical area, i.e. $\Gamma A/A_{geom} = \Gamma r$. The resulting values are listed in Table 6.2, and give an indication of the amount of protein deposited in each electrode. As anticipated, the higher Mb surface coverage on the CNT composite as compared to the other carbon electrodes is probably due to its larger porosity and roughness degree, which facilitates the immobilization of higher amounts of protein and amplifying the signal response.

6.1.2 Electrocatalytic Activity

We have proved that direct electron communication is possible between the active redox centre of the immobilized protein and the carbon electrode. We have concluded that carbon nanotube materials, either as CNT forest or as CNT composites, exhibit improved electrochemical performance than the graphite composites. In the case of the forest CNT, increased electron transfer kinetics is achieved whether in the case of the CNT composites, their high degree of porosity facilitates the immobilization of more amount of protein and therefore the electrochemical signal is amplified.

Then, a question has been raised: besides of having good electron communication, does the protein still preserves its bioactivity to typical analytes (O_2 and H_2O_2) when anchored to the different carbon electrodes? Subsequently, we have focused on two cases: Electrocatalysis of oxygen and Electrocatalysis of hydrogen peroxide.

6.1.2.1 Electrocatalysis of Oxygen

In anaerobic conditions, the electrochemical response of the heme proteins is characterized by the appearance of a well defined and reversible pair of redox peaks (Protein-Fe(III)/Protein-Fe(II)) as a consequence of the good electronic coupling between the heme core and the electrode surface. However, when the metalloproteins are exposed to oxygen, a strong catalytic effect appears. This can be observed in Fig. 6.4 which compares the electrochemical behaviour of the reversible Fe(III)/Fe(II) redox pair corresponding to two different heme proteins (myoglobin and catalase) when immobilized on forest-like CNT electrodes (taken as a reference) either in oxygen-free PBS solution or in air saturated buffers. In the latter case an electrocatalytic effect can be observed which is reflected in an increase of the Fe(III) reduction peak accompanied by the disappearance of Fe(II) oxidation peak [2, 8, 9, 13–15].

The mechanism of the catalytic reaction is still not completely understood but the more plausible mechanism is the following: as the potential is cathodically scanned, the direct electrochemical reduction of Protein-Fe(III) to Protein-Fe(II) takes place at the carbon electrode. Fe(II) has a very high affinity for oxygen and the latter process is immediately followed by a fast reaction of Protein-Fe(II) with oxygen. The product of Protein-Fe(II)–O_2 could then undergo again electrochemical reduction at a potential close to Protein-Fe(III) reduction, producing H_2O_2 and Protein-Fe(II). Protein-Fe(II) again reacts fast with O_2 to continue the cycle. This scheme of reactions is consistent with the large cathodic wave (due to the catalytic

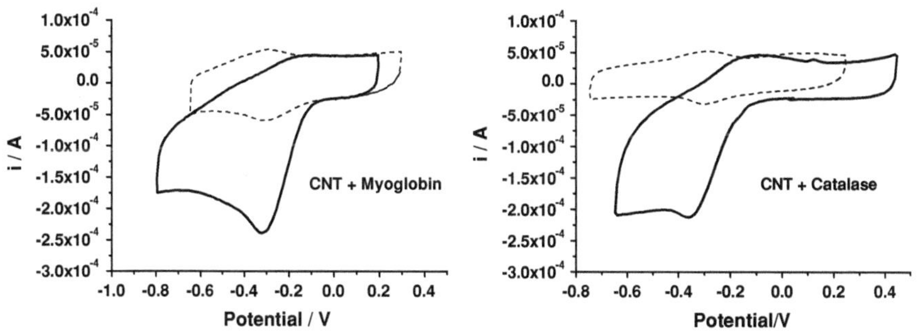

Fig. 6.4 Cyclic voltammograms for a forest-like CNT electrode modified with myoglobin and catalase in deareated PBS solutions (*dashed curve*) and air saturated PBS solutions (*solid curve*). Scan rate: 0.1 V/s

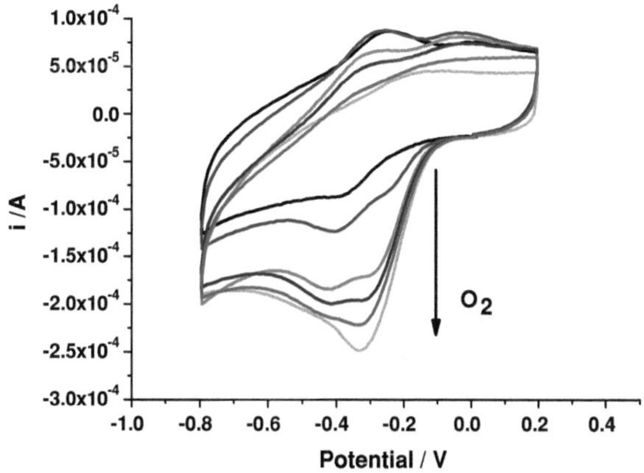

Fig. 6.5 Cyclic voltammetry of Mb/CNT forest electrodes in presence of increasing oxygen amounts in PBS solutions. *Black solid line* represents the electrochemical behaviour of the electrode in N_2 saturated solution. In the following curves the electrochemical system was bubbled with oxygen during 1, 5, 8, 11 and 13 min. Scan rate of 0.1 V/s

reduction of Fe(II)–O_2) and the disappearance of the Fe(II) anodic wave (due to the fast reaction of Fe(II) with oxygen) [16–18].

$$\text{Protein-Fe(III)} + e^- \rightarrow \text{Protein-Fe(II) (at electrode)}$$
$$\text{Protein-Fe(II)} + O_2 \rightarrow \text{Protein-Fe(II)}-O_2 \text{ (fast)}$$
$$\text{Protein-Fe(II)}-O_2 + 2H^+ + 2e^- \rightarrow \text{Protein-Fe(II)} + H_2O_2 \text{ (at electrode)}$$

Figure 6.5 illustrates the development of catalytic peaks in the presence of limiting oxygen reactant at pH = 7.3. As starting point, the Mb response in oxygen-free PBS solution can be taken as a reference. In this case, the pair redox couple MbFe(III)/MbFe(II) appears at a formal potential of 0.314 V at scan rate of 100 mV/s. The remaining curves correspond to the cyclic voltammograms

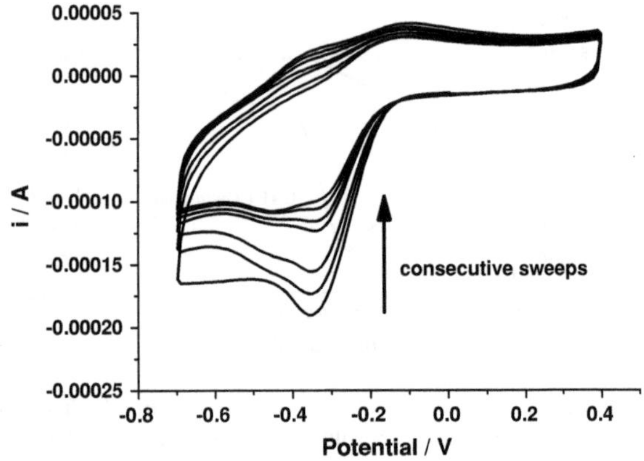

Fig. 6.6 Electrochemical response of CNT/Mb in aereated PBS solutions at different successive scan cycles. Scan rate: 0.1 V/s

containing increasing oxygen concentrations. An extra cathodic peak appears which increases with the oxygen amount up to dominate the overall cathodic wave. The extra peak can be associated to the electrochemical reduction of MbFe(II)–O_2 which can undergo reduction at slightly more positive potentials than MbFe(III)/MbFe(II), consistent with a fast electrochemical catalysis [2, 13].

Another way to observe the oxygen electrocatalytical effect of myoglobin but in a reverse way than in Fig. 6.5 is by performing successive scans on the Mb/CNT system (Fig. 6.6). In the first scan the typical response of the MbFe(III)–O_2 reduction is observed reflected by a large cathodic wave. With consecutive scans, the cathodic wave decreases with the consequent appearance of two peaks corresponding to the reduction process of MbFe(III) and MbFe(II)–O_2, with the former one appearing at more cathodic potentials. These results suggest that in the time scale of the consecutive runs progressive oxygen depletion is built up in the diffusion layer, which difficults the oxygen uptake by the immobilized protein when Mb Fe(II) is cathodically formed at the electrode. If the scanning is stopped and a new i/V run is performed, the electrocatalytic oxygen reduction process on Mb is recovered.

Electrochemical Response on Different Substrates

In Fig. 6.7, a comparison of the electrochemical response of the different Mb-modified carbon electrodes in an aerated solution is illustrated. Figure 6.4 also shows, as a reference, the voltammograms of protein-free carbon electrodes, where there was no signature of a peak, at least in the scanned potential window. It can be observed that in terms of current density, the CNT composite exhibits the higher values, followed by the forest-like CNTs, and the graphite composite. The reason for that has to do with the protein coverage on the different carbon-based systems, as discussed previously. Moreover, the sensitive response of Mb to oxygen also demonstrates that the immobilization procedure on such carbon systems does not perturb the bioactivity of the protein.

Another set of electrode comparisons can be observed in Fig. 6.8 which shows the oxygen/Mb response on HOPG, gold electrodes and CNT forest surfaces

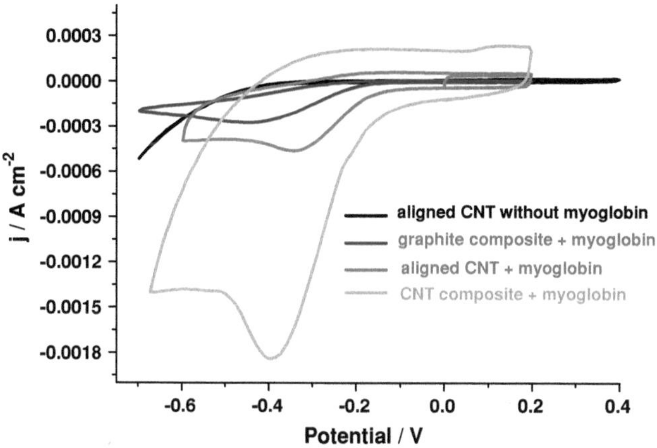

Fig. 6.7 i/V profiles for the different carbon electrodes in aerated PBS solutions at a scan rate of 0.1 V/s

normalized to the same geometric area. Three voltammogram scans are depicted for each electrode material; they were taken with 3 min of difference among them. It is remarkable the poor voltammetric response achieved in the case of HOPG and Au electrodes. The oxygen catalytic response progressively disappeared in different i/V runs in the case of Au and HOPG which can reflect either difficulties in anchoring firmly the protein on such electrode surfaces or a progressive denaturation of the protein with the consequent loss of its bioactivity. However, a more stable and reproducible response was kept at the forest CNT substrates. In the results presented in Fig. 6.8b and c, Mb has been physisorbed on HOPG and Au electrodes whereas in the case of the CNT forest, a covalent attachment of Mb to the CNT was carried out. However, it is worth mentioning that physisorbed Mb on CNT forest was also studied leading to similar conclusions than in the case of covalent attachment. The main difference between the system with a higher degree of covalently attached protein and the one with only physisorbed biomolecules could be reflected in the long term electrode lifetime. We verified that covalently attached Mb/CNT systems could provide reproducible response up to a month with only a decrease of the electroactivity of about 15 %, whereas Mb physisorbed on CNT electrodes exhibited stable electroactivity response only for a maximum of one week. During this time and when they were not in use, the electrodes were stored in PBS solutions at 4 °C.

The Effect of pH

The pH modulates not only the net charge density of the protein but also the accessibility of water to the Mb pocket and protonation of the heme surroundings. Such effects can induce protein conformational changes and alter its bioactivity. Therefore, the influence of pH on the electrochemical oxygen reduction was

Fig. 6.8 Oxygen reduction catalytical signal for different electrode surfaces in PBS solutions at 0.1 V/s. Three cyclic repetitive runs are shown for each electrode taken with a difference of 3 min among them (*black line* 1st cycle; *grey* 2nd cycle; *light grey* 3rd cycle)

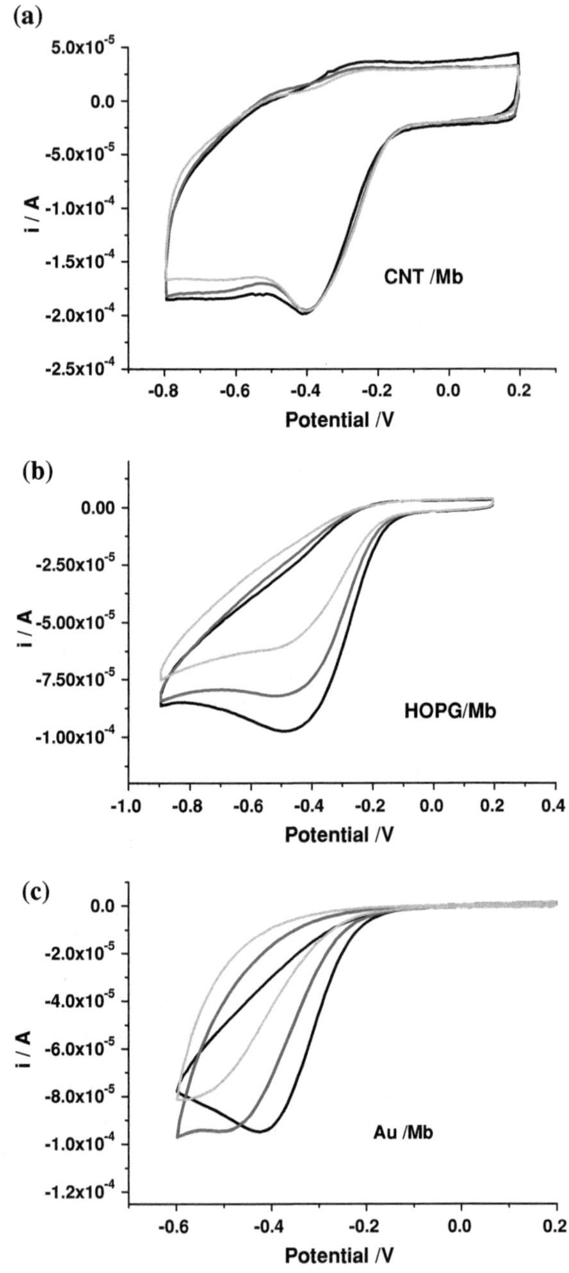

also investigated on Mb/CNT modified electrodes in a pH range from 5.0 to 9.4. As shown in Fig. 6.9a, the peak potentials of Mb/CNT-forest modified electrodes shifted negatively with increasing pH. The dependence of the peak potential on pH

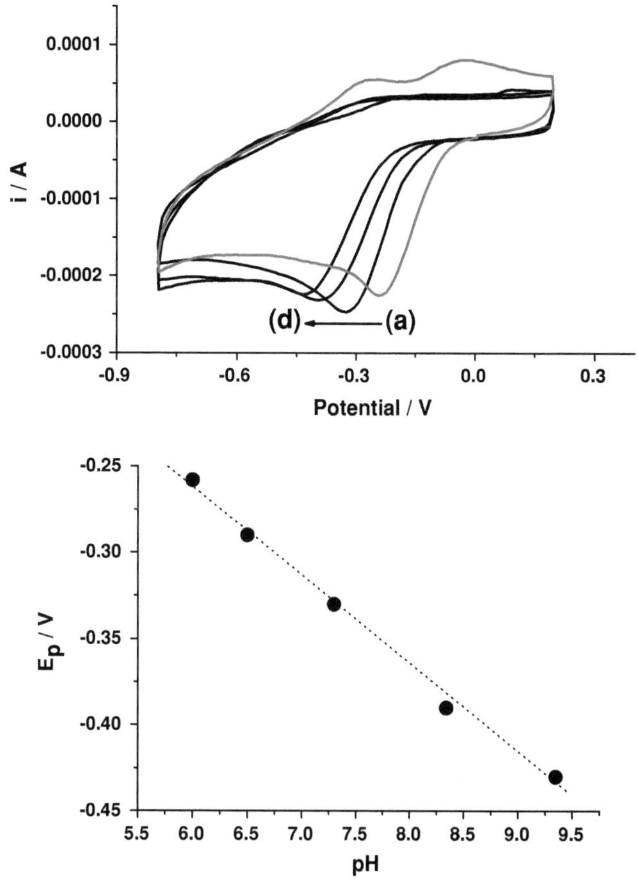

Fig. 6.9 Cyclic voltammograms of CNT/Mb in PBS at different pH: 6.0 (**a**), 7.3 (**b**), 8.34 (**c**), 9.35 (**d**). **b** pH = 5.4; and plot of E_p versus pH. Scan rate: 0.1 V/s

showed a linear relationship in the range of pH 6–9.4 with a slope of −52 mV/pH with a correlation coefficient of 0.993 (Fig. 6.9b). This feature clearly indicates that the Mb undergoes proton-coupled electron transfer.

However, for values of pH below 6.0 (Fig. 6.10), at which the protein was positively charged, we observed important, reproducible and reversible changes in its cyclic voltammogram characteristics.

It seems that the protein undergoes conformational changes as the pH is decreased which is represented by the appearance of more anodic electroactivity. The electrochemical response stabilizes in two reversible well defined redox peaks (MbFe(II)/MbFe(III)) and the electrocatalytical oxygen wave disappears. This behaviour suggests that under such low pH, the protein is unable to bind oxygen. Conformational changes associated with a partly unfolded deoxyhemoglobin have been reported at low pH by monitoring the electrochemical and visible/

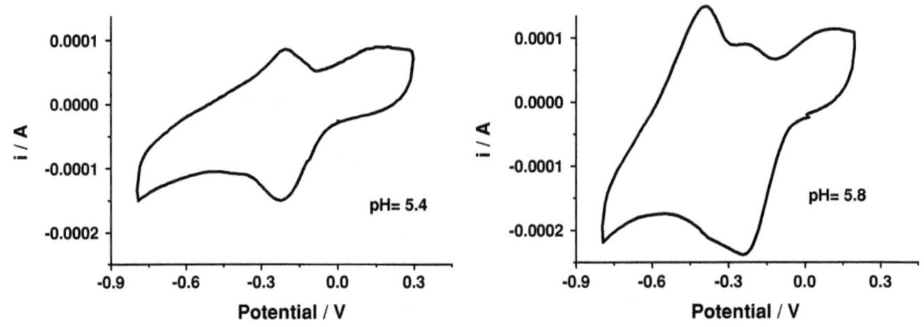

Fig. 6.10 Cyclic voltammograms of Mb/CNT-forest in aerated PBS at low pH (pH = 5.4 and 5.8). Scan rate: 0.1 V/s

infrared spectroscopy response [2, 7]. However, no studies have been carried out by following pH dependence on the oxygen binding protein activity, which is of relevant importance for sensor technology.

6.1.2.2 Electrocatalysis of Hydrogen Peroxide

As mentioned previously, heme proteins also exhibit catalytic activity to hydrogen peroxide when immobilized on carbon-based electrodes. When aliquots of H_2O_2 were added to the PBS solution, a considerable increase of the reduction peak was observed at about the same potential at which the electrocatalytic oxygen reduction was taken place. The disappearance of the anodic peak was also detected. All these features can be observed in Fig. 6.11, in which the behaviour of the vertically aligned CNT electrode is taken as a representative example. A reference test using the carbon electrodes in absence of protein but in presence of different concentrations of H_2O_2 showed no obvious cathodic peak response, at least in the potential range under study.

The catalytic reduction of H_2O_2 exhibits a behaviour similar to the one found with oxygen, which suggests that the reaction mechanisms could be similar. Several plausible explanations have been provided in the literature [16–18], under the assumption that H_2O_2 can act as an oxidant or reductant agent. It has been suggested that H_2O_2 reacts with MbFe(III) to give an oxyferryl radical (•MbFe(IV) = O) which can react again with H_2O_2 to produce oxygen and Mb(III). When MbFe(III) is reduced at the electrode, it can react fast with oxygen and follow the same pathway described in the previous section for the oxygen.

Therefore, one of the pathways suggested in the literature is:

$$\mathrm{MbFe(III)} + \mathrm{H_2O_2} \rightarrow {}^{\bullet}\mathrm{MbFe(IV)} = \mathrm{O} + \mathrm{H_2O}$$
$${}^{\bullet}\mathrm{MbFe(IV)} = \mathrm{O} + \mathrm{H_2O_2} \rightarrow \mathrm{MbFe(III)} + 1/2\mathrm{O_2} + \mathrm{H_2O}$$
$$\mathrm{MbFe(III)} + \mathrm{e^-} \rightarrow \mathrm{MbFe(II)}\ \text{(at electrode)}$$
$$\mathrm{MbFe(II)} + \mathrm{O_2} \rightarrow \mathrm{MbFe(II)}-\mathrm{O_2}$$
$$\mathrm{MbFe(II)}-\mathrm{O_2} + 2\,\mathrm{e^-} + 2\,\mathrm{H^+} \rightarrow \mathrm{MbFe\,(II)} + \mathrm{H_2O_2}\ \text{(at electrode)}$$

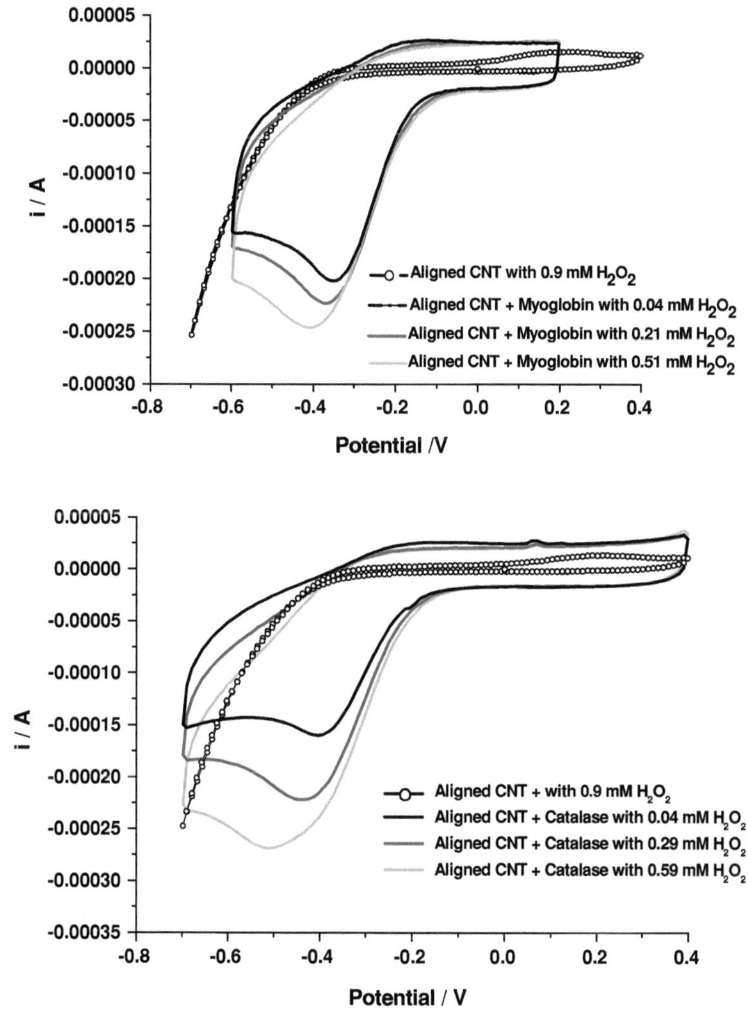

Fig. 6.11 i/V voltammograms for the vertically aligned CNT electrode modified with Mb or catalase in PBS solutions under different concentrations of H_2O_2 at a scan rate of 0.1 V/s. The electrochemical response to H_2O_2 of a similar electrode before the protein immobilization is also included

However, from the electroanalytical point of view, the different carbon electrodes do not behave in the same way with the adding of H_2O_2 aliquots. For a closer inspection of the electrochemical performance of each electrode, from Figs. 6.12, 6.13, 6.14 represent the change in the peak current density as a function of the peroxide concentration for the three carbon electrodes taking myoglobin as representative system. In the case of the graphite composites (Fig. 6.12), the results clearly show that the reduction peak current increases with an increasing concentration of H_2O_2 above 1×10^{-4} M but only keeps a linear behaviour for a reduced range of

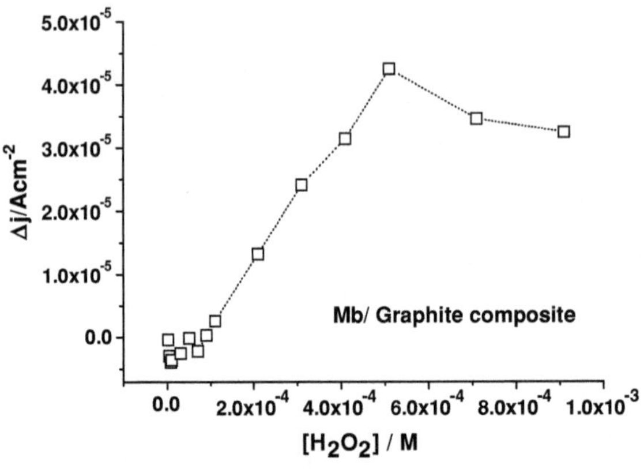

Fig. 6.12 Current change as a function of H_2O_2 concentration for the myoglobin modified graphite composite

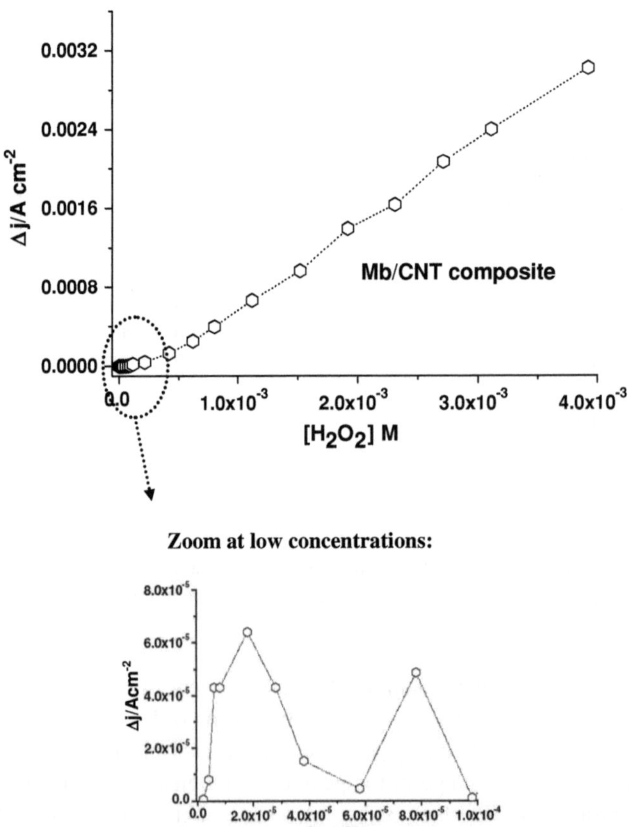

Fig. 6.13 Current change as a function of H_2O_2 concentration for the myoglobin modified CNT composite electrode. For a closer inspection, a zoom at low H_2O_2 concentrations is also shown

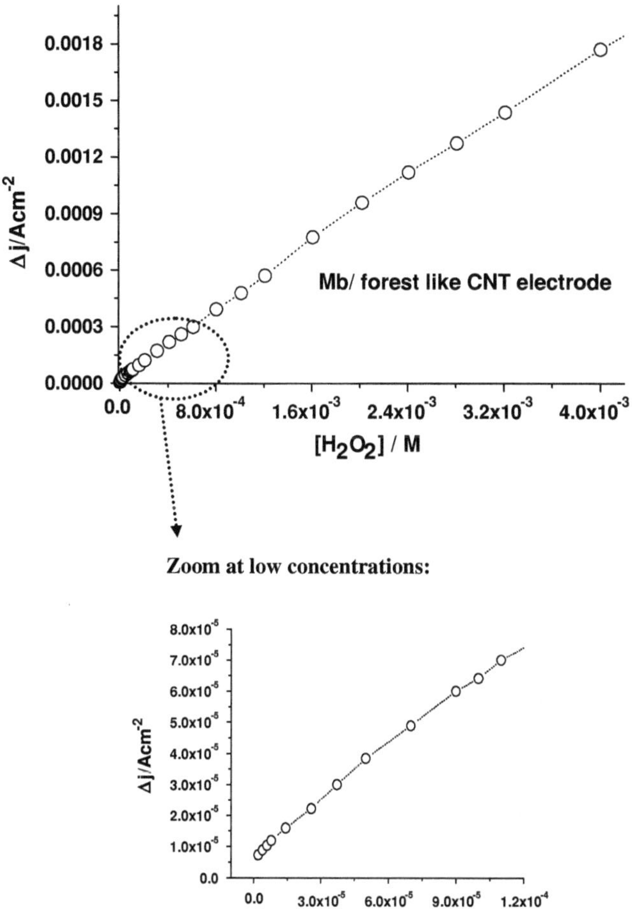

Fig. 6.14 Current change as a function of H_2O_2 concentration for the myoglobin modified vertically aligned CNT electrode. For a closer inspection, a zoom at low H_2O_2 concentrations is also shown

concentrations. Above a certain concentration, it seems that the electrode reached saturation and upon further addition of H_2O_2, the current density slightly decreases probably due to a partial protein inactivation in the presence of a higher concentration of peroxide. In the linear range, the slope of the curve, which gives us information of the sensitivity of the sensor system, was approximately 0.14 A cm^{-2}/M.

In the case of the carbon nanotube composite, a satisfactory calibration curve could be obtained above 2×10^{-4} M and the linear behaviour was extended in a wide range of concentrations with a remarkable high sensitivity of about 0.84 A cm^{-2}/M. The zoom in Fig. 6.13 shows that the signal is very noisy for concentrations below 1×10^{-4} M, which impedes an accurate detection.

However, very low limits of detection were achieved with the vertically aligned carbon nanotube electrode (Fig. 6.14). In these electrodes, a monotonic behaviour

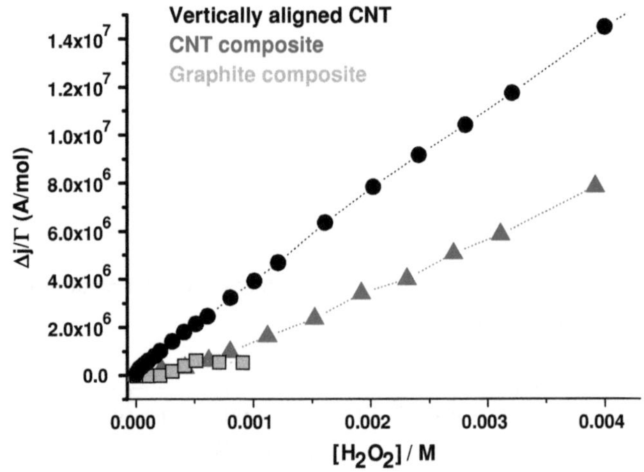

Fig. 6.15 Current density normalized with respect to the protein coverage as a function of H_2O_2 concentration for the different myoglobin modified carbon electrodes

Table 6.3 Electroanalytical parameters for the detection of hydrogen peroxide on the different carbon-based electrodes

Electrode	LOD	Sensitivity (A cm^{-2} M^{-1})	Linear range
Graphite composite	$9{\cdot}10^{-5}$ M	0.14	Small
CNT composite	$6{\cdot}10^{-5}$ M	0.84	Wide
Forest CNT	$5{\cdot}10^{-8}$ M	0.47	Wide

The LOD values were estimated as three times the noise level of the blank

of the current as a function of H_2O_2 concentration was observed in a wide range of concentrations and with an average slope of 0.47 A cm^{-2}/M (Fig. 6.14).

Among the different analysed electrode systems, it is remarkable the high signal/noise ratio of the vertically aligned CNT electrode at such low H_2O_2 concentration. This could be as consequence of a facilitated electron transfer process of the protein in such environment. Higher electron transfer rates between the electrode and the protein redox centre, as in the case of the forest-like CNTs, can enhance the protein turnover and favour the electrocatalytical process.

In order to have a better quantification of the sensitivity of the different electrodes to the analyte, a normalization of the curves with respect to their electrochemically active protein coverage was made and the results are shown in Fig. 6.15. At the same electroactive protein coverage, the vertically aligned CNT electrode is almost two times more sensitive than for the case of the CNT composite. Thus, the higher apparent sensitivity of the CNT composite electrode can be only ascribed to the larger protein coverage due to the higher degree of porosity of this electrode. On the other side, the graphite composite shows a sensitivity similar to that of the CNT composite but exhibits the serious drawback of a poor range in linear behaviour with increasing analyte concentration.

Table 6.3 summarizes the electroanalytical characteristics of the myoglobin graphite composites, CNT composites and CNT forest electrodes.

These results highlight the good electrochemical performance of electrodes made by a vertical disposition of CNT. The reasons underlying this enhanced performance can be attributed to the fact that this particular configuration results in a more suitable environment for the covalent anchoring of the protein, providing a favourable orientation of the proteins on the tubular structure while keeping a robust protein immobilization. At the same time, the vertical arrangement of CNT facilitates a high surface density of CNT edges, which have been shown to exhibit an improved and fast electron transfer response. All these features promote a better electron coupling with the protein system which can result in a faster electrode response.

6.1.3 General Conclusions of Iron Protein-Based Amperometric Biosensors

So far, we have demonstrated that carbon nanotubes electrodes provide an optimum environment for facilitating direct electron transfer with heme proteins and for keeping active the protein function to target analytes (O_2 and H_2O_2). It has also been shown that CNT composites provide an interesting platform with high surface area for biomolecules accumulation and current response amplification. Even more remarkable, forest-like carbon nanotube electrodes provide a more facilitated electron transfer which is reflected in improved electroanalytical characteristics such as higher sensitivity, wider analyte concentration range and lower limits of detection.

6.2 Direct (Label-Free) Electrochemical DNA-Detection

With respect to DNA-detection, it is very common to find in literature many procedures relying on chemical labelling of target DNA sequences [19–24]. However, as mentioned in Chap. 1, the extra labelling step either in the target DNA or in the synthetic oligonucleotide makes the process more expensive and effortful.

In this context, we would like to pursue DNA-detection but focusing on more direct or label-free detection methods. Therefore, in this section we will evaluate the use of CNT electrodes for such purposes. We will start addressing the DNA base oxidation as detection scheme on CNT electrodes, highlighting the advantages and disadvantages of the method together with the benefits that can provide the use of CNTs.

In the second part, we will shift to the detection of hybridization events through also a label-free strategy which combines several ingredients for providing an easy electrochemical detection procedure. On one side, we will use high density vertically aligned CNT microelectrode arrays whose performance will be illustrated by establishing comparisons with glassy carbon modified with randomly dispersed CNTs or macroscopic CNT forest electrodes. On the other side, we will use highly positively charged redox reporters as indicators of the hybridization process, which constitutes an interesting alternative to the chemical labelling DNA-detection approach. The biorecognition event will be followed up by cyclic voltammetry, chronocoulometry and impedance spectroscopy. The DNA-detection scheme will profit from the CNT versatility to develop new miniaturized electrode configurations and also from the CNT suitability to chemical DNA wiring. Moreover, we will show that the scaling down of the CNT electrode size (CNT microelectrodes) results crucial for the success of such practical and simple DNA-detection strategy.

6.2.1 Electrochemical Detection of DNA by Oxidation of the DNA Bases

The simplest strategy of label-free detection of DNA or hybridization event is based on the direct oxidation of the DNA bases physisorbed on the electrode material. As mentioned in Chap. 1, it greatly simplifies the detection saving time and cost of the assay. However, this direct strategy suffers from various drawbacks as detailed in the introductory part. Some of them can be tackled by using CNT electrodes. One of the problems is related with the base oxidation potential. DNA bases exhibit high oxidation potentials and consequently higher background currents which can mask he electrochemical signal [25, 26]. Among the DNA bases, guanine is the one which exhibits the lowest electrochemical overpotential for oxidation, followed by adenine, but still the required value is relative high (close to 1 V). CNTs can help to increase the CNT signal response from the background noise due to surface area effects (more loading of the probe and target species). In some cases, it has been also reported a decrease of the oxidation overpotentials when using CNT systems [21, 25, 26]. Motivated by these promising features and especially on the possible overpotential decrease, we followed up the DNA oxidation in novel upright CNT microelectrodes and also on glassy carbon electrodes modified by randomly dispersed CNTs as a matter of comparison. The electrochemical response of Poly-G physically adsorbed on CNT microelectrodes, on CNT modified glassy carbon and on glassy carbon was followed by differential pulse voltammetry (DPV) which allowed us to subtract the electrolyte background response and to have a better redox guanine signal. As a matter of fact, Fig. 6.16a shows the DPV curves for a glassy carbon and a CNT modified glassy carbon electrode. We observed that the presence of the CNTs remarkably increase the guanine oxidation current due to the increased adsorption of Poly-G provided by the higher surface area of the CNTs. Well defined DPV profiles were also observed in the case of CNT microelectrodes as shown in Fig. 6.16b. In all CNT tested electrodes, the guanine oxidation potential was about 0.9 V (vs. Ag/AgCl), just the same value found when guanine oxidizes on the glassy carbon surface.

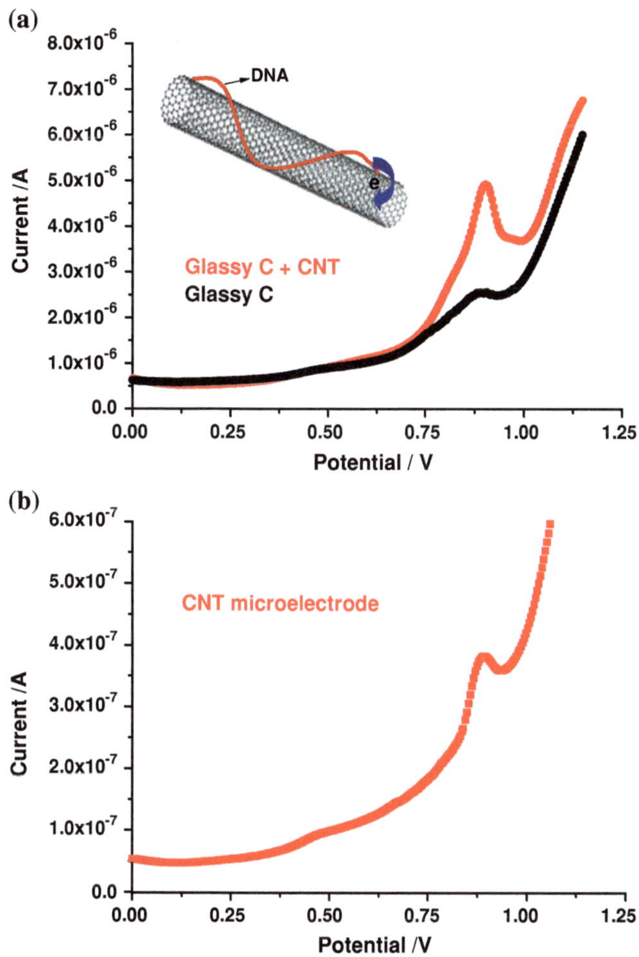

Fig. 6.16 **a** Differential pulse voltammetry of a glassy carbon/CNT electrode with physisorbed Polyguanine (Poly-G) (in *red*) and of a glassy carbon electrode with physisorbed Poly-G (in *black*). The peak corresponds to the direct oxidation of guanine. Note the enhanced signal of the Poly-G oxidation when it is adsorbed on the CNT support. That is due to the high surface area of CNTs which allow more analyte loading on their surface. **b** DPV of CNT microelectrode (Color Figure Online)

That indicates that our CNT systems are not inducing any decrease in the redox overpotential and in our case, the only positive contribution of using CNTs is the enhanced electrochemical signals due to surface area effects.

The absence of an overpotential decrease for DNA base oxidation with our CNT platforms together with the additional drawbacks mentioned in Chap. 1 discouraged us from continuing developing such strategy for DNA-sensing. Under these perspectives, we have shifted to other DNA-detection approaches that can profit of simplicity but also of practicality.

6.2.2 *Electrochemical Detection of DNA Hybridization by Using Reversible Redox Indicators*

Another direct method for following up the biorecognition process is to use redox active reporter molecules that intrinsically associate with the double helix in a non-covalent manner. That approach avoids the sample damage that takes place when using base oxidation or the difficulties of using covalently attached redox labels in the DNA. These reporter molecules can interact with the DNA either by electrostatic interactions or by intercalative means (in such cases, DNA has been considered by some authors as a charge mediator) [27, 28]. Some examples of the former molecules are the highly positively charged cobalt and ruthenium complexes [i.e. $Co(phen)_3{}^{3+}$, $Ru(NH_3)_6{}^{3+}$]. These positively charged complexes can strongly interact with the negatively phosphate backbone. As the hydridization process takes place, proportionally more positively charged complexes bind DNA yielding to a higher signal.

The other kind of reporter molecules comprises redox probes that can be specifically intercalated in double strand DNA. An example of these sorts of molecules is the methlylene blue. These molecules have also been used as reporters on perturbations in base stacking.

The approaches with non-covalent and reversible redox reporters are highly sensitive, simple and do not damage the sample which gives the advantage of further use of the sample. Moreover, the redox molecule reporter strategy together with the functionalization and signal amplification capabilities of the CNTs, the powerful electrochemical detection techniques and the possibilities of sensor miniaturization can provide important stimulus to the field of DNA-detection and diagnosis.

In this work $Ru(NH_3)_6{}^{2+/3+}$ has been used to signal DNA hybridization on CNT microelectrodes on glassy carbon/CNT electrodes and on macroscopic forest CNT electrodes to establish comparisons. Such ruthenium complexes are cleaner species which do not adsorb on the carbon electrodes as compared to other electrostatic indicators [$(Fe(CN)_6{}^{3-/4-}$] or intercalators (e.g. daunomycin) which can smear the signal coming from the interaction with the DNA. Oligomer Poly-C-NH_2, used as probe, was covalently attached to the CNTs through carbodiimide chemistry, then the surface was blocked with PEG for non-specific adsorption minimization and after that, Poly-G was used as target. The $Ru(NH_3)_6{}^{2+/3+}$ electrochemical signal was monitored before and after the hybridization process by cyclic voltammetry, chronocoulometry and impedance measurements. The scaling down of the electrode size brings about a change in the control regime of the electrochemical process which is particular beneficial for the impedimetric detection under the chosen redox reporter protocol; that procedure would have not been possible in macroscopic CNT electrodes, but this will be explained in detail, later on. Moreover, chronocoulometric measurements can provide more quantitative information of the immobilized DNA probe and of the hybridization the efficiency.

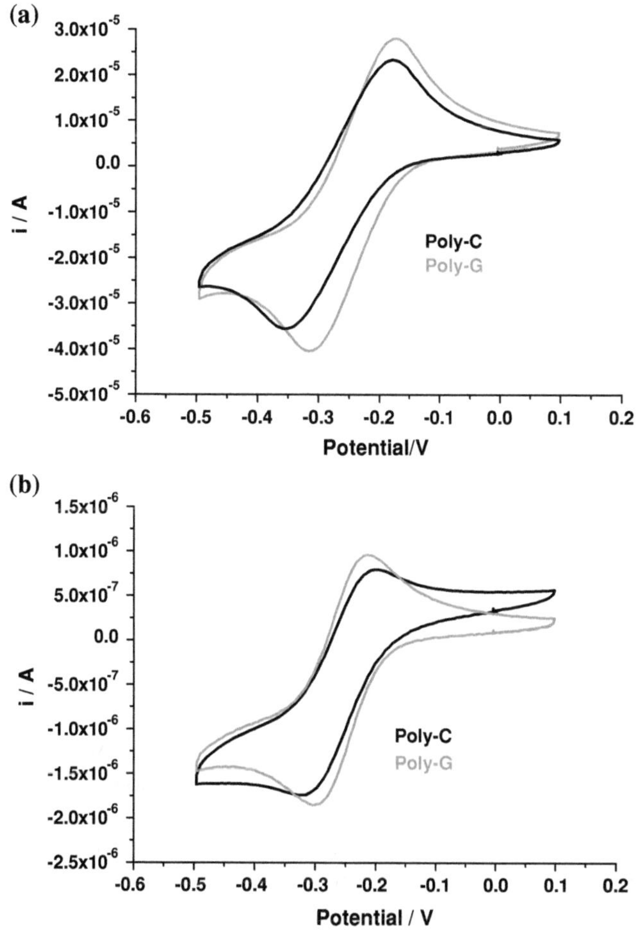

Fig. 6.17 Cyclic voltammograms for **a** glassy carbon/CNT electrode and **b** CNT microelectrode modified with covalently attached Poly-C and after immersion in $1 \cdot 10^{-6}$ M Poly-G. The electrodes were blocked with PEG to avoid on-specific adsorption. The electrolyte was 1.8 mM $Ru(NH_3)_6{}^{3+}$ prepared in TRIS buffer solution and was deoxygenated with N_2 bubbling for 5 min prior to measurements

6.2.2.1 Cyclic Voltammetry

Cyclic voltammetry (CV) is a fast approach to detect the hybridization event. Although it does not exhibit the high sensitivity of the other two techniques that will be shown later on, it is very suitable for providing the general trend in the hybridization process.

Figure 6.17 shows the CV response of glassy carbon/CNT electrode and CNT microelectrode array modified with covalently attached Poly-C probe and after

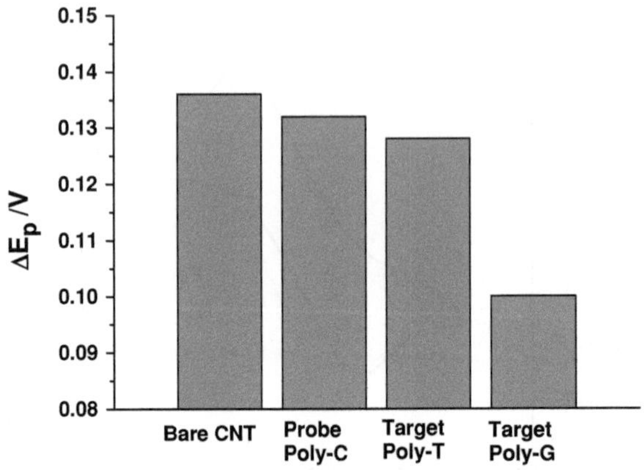

Fig. 6.18 Peak potential difference histogram of bare CNT microelectrode, CNT microelectrode modified with Poly-C and when is targeted with complementary $1 \cdot 10^{-6}$ M Poly-G and with non-complementary 1.10^{-6} M Poly-T. The electrode was blocked with PEG for avoiding non-specific adsorption. The electrolyte was 1.8 mM $Ru(NH_3)_6{}^{3+}$ prepared in TRIS buffer solution and was deoxygenated with N_2 bubbling for 5 min prior to measurements

interaction with 1×10^{-6} M Poly-G target. An augment of the peak current and a decrease in the peak potential difference are observed when hybridization takes place. Such features indicate an electrostatic surface accumulation of the cationic redox marker due to the increase of the negatively charged oligonucleotide phosphates with the increasing hybridization. Such accumulation of the cationic redox marker at the electrode interface makes electron transfer kinetics of the redox probe more favourable which is reflected in the peak potential difference decrease.

A better inspection of this electrochemical response can be followed in Fig. 6.18 where the selectivity of the system is proved. The figure shows the peak potential differences for the bare CNT microelectrode array, for the same electrode covalently modified with Poly-C probe and when is targeted with the complementary and non-complementary DNA strand; the peak potential differences are smaller when the target is the complementary strand.

Once the selectivity of the system has been proved, the sensitivity should be tested. CNT microelectrode arrays modified with Poly-C were immersed in increasing concentrations of Poly-G strands. Figure 6.19 shows a detailed plot of the change of the peak potential difference of the Poly-C/CNT microelectrode in presence of increasing aliquots of target oligonucleotides.

It is worthy to mention that these features were also found in the glassy carbon/CNT platform. A histogram of the peak potential difference as a function of increasing target DNA concentrations is shown in Fig. 6.20.

From all the results it can be clearly confirmed the trend previously observed: an augment of the peak current and a decrease in the peak potential difference as the concentration of the target DNA is increased. Such features indicate the accumulation of the cationic redox marker by electrostatic interaction with the negatively charged DNA and a consequently enhancement of the electron transfer kinetics of

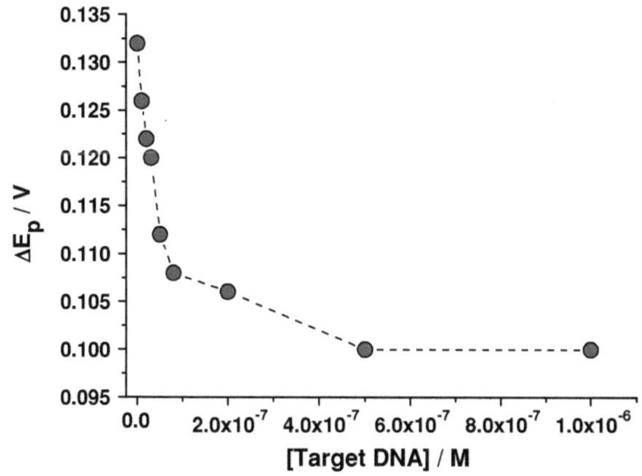

Fig. 6.19 Plot of the peak potential difference as a function of increasing target DNA concentrations. The electrode was blocked with PEG for avoiding non-specific adsorption. The electrolyte was 1.8 mM $Ru(NH_3)_6^{3+}$ prepared in TRIS buffer solution and was deoxygenated with N_2 bubbling for 5 min prior to measurements

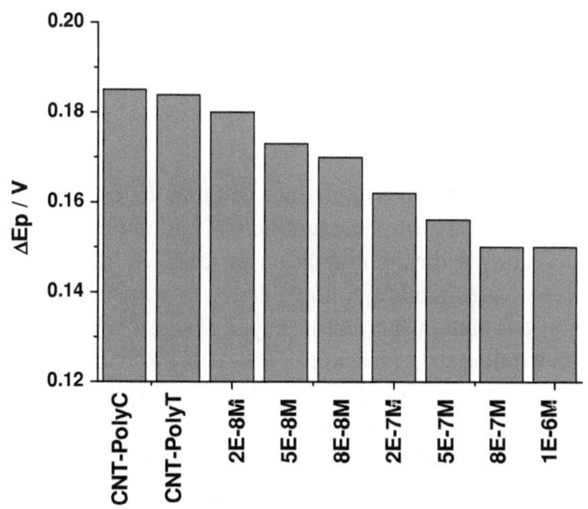

Fig. 6.20 Peak potential difference histogram of glassy carbon/CNT modified with covalently attached Poly-C in presence of non-complementary $1 \cdot 10^{-6}$ M poly-T and in presence of increasing of complementary Poly-G targets. The electrode was blocked with PEG for avoiding non-specific adsorption. The electrolyte was 1.8 mM $Ru(NH_3)_6^{3+}$ prepared in TRIS buffer solution and was deoxygenated with N_2 bubbling for 5 min prior to measurements

the redox marker. These results also demonstrate that such CNT platforms are able to anchor the DNA probe and follow the hybridization process. However, this technique can not discriminate between the redox response of the ruthenium species diffusing from the solution and those ones electrostatically adsorbed on the phosphate backbone when both of them are present simultaneously. For discrimination of both kinds of species, chronocolumetric measurements can be performed.

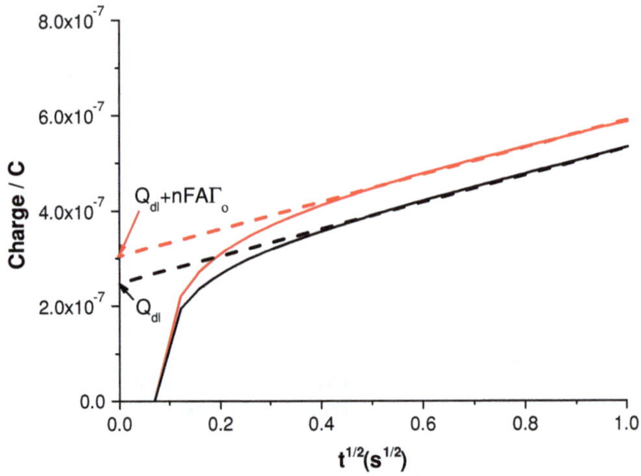

Fig. 6.21 Illustration of the procedure in order to get the amount of electroactive species adsorbed on a CNT/Poly-C electrode. First the chronocoulometric measurements were performed in the background electrolyte, buffer solution of 10 mM TRIS + 10 mM NaCl (*black curve*) in order to extract the Q_{dl}. Then the solution was changed to 50 μM of $Ru(NH_3)_6^{3+}$ in TRIS + NaCl in order to determine the amount of the adsorbed redox species (*red curve*). Potential pulse from 0.1 to −0.4 V and duration time of 0.5 s

6.2.2.2 Chronocoulometry

We used chronocoulometry measurements in order to estimate the accumulated cationic redox marker at the oligonucleotide CNT modified surface and take such parameter as indicator of the hybridization process. As described in Chap. 1, to estimate the charge corresponding to the electroactive species adsorbed on the surface, one has first to subtract the double layer charge [29]. Accordingly, we have performed chronocoulometric measurements with the background electrolyte (in absence of the electroactive species that undergo adsorption on the electrode). By extrapolation of a least squares fit from the linear part of the curve charge versus $t^{1/2}$, one can obtain the double layer charge (Q_{dl}) from the intercept at time zero. Then the chronocoulometric measurements are performed in presence of the redox markers and again the interception at zero time is estimated which corresponds to the charge of the double layer plus the charge of the adsorbed amount. Therefore, from both measurements one can get the adsorbed amount by Q_{dl} subtraction (Fig. 6.21). As compared to CV, the chronocoulometry has the advantage that the double layer charge and the charge due to the redox reaction of species adsorbed on the electrode surface can be differentiated from the charge due to the redox reaction of the molecules that diffuse to the electrode. That avoids confusion over which concentration is being measured, thus simplifying the analysis.

Such methodology has been followed to calculate the charge corresponding to the amount of electroactive species adsorbed on the electrode ($Q_{ads} = nFA\Gamma_0$) and to monitor the hybridization process [29]. In the low ionic strength buffer solution we

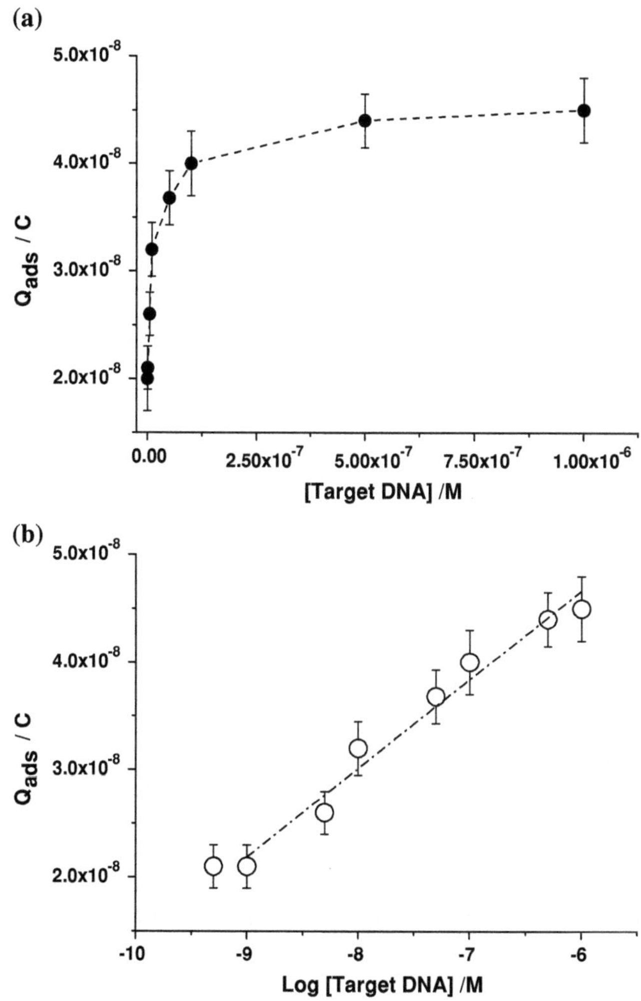

Fig. 6.22 Q_{ads} on a CNT microelectrode array chemically modified with Poly-C **a** as a function of increasing concentrations of target Poly-G, **b** as a function of the logarithm of increasing concentrations of Poly-G. *Error bars* correspond to triplicate measurements

are using, the trivalent ruthenium species preferentially exchanges with the native monovalent DNA counterion and associates with the phosphate groups in a ratio 1:3.

Figure 6.22 shows the Q_{ads} as a function of increasing aliquots of target DNA for the CNT microelectrode. It can be observed that the probe modified electrode is very sensitive to concentrations below 2.5×10^{-7} M of the target and then saturation is reached. The limit of detection in such CNT microelectrode system was found to be around 1 nM (calculated from three times the standard deviation of the blank). Controls with non-complementary strands showed no increase in the charge associated with the electrode surface.

The use of this technical tool in combination with CNT platforms has demonstrated to be a quite sensitive strategy to monitor the hybridization process.

One can also go a step forward and calculate Γ_o, the amount of adsorbed redox marker confined near the electrode, from Q_{ads} to get more details of the hybridization process [29, 30]. In order to calculate Γ_o, we need to estimate the electrochemical active area of the microelectrode array. For that, we have used the peak current from a cyclic voltammogram of the bare CNT microelectrode in presence of the ruthenium species. By applying the Randles–Sevcik equation:

$$i_{\text{peak}} = 2.69 \times 10^5\, n^{3/2} D^{1/2} C_{\text{bulk}} \text{v}^{1/2} A$$

where n is the number of electrons involved in the redox process, D is the diffusion coefficient of the ruthenium species ($D = 9 \times 10^{-6} \text{cm}^2/\text{s}$), C_{bulk} is the concentration of the ruthenium complex, v is scan rate and A is the surface area. From such equation we have got an electrochemical active area of 3.9×10^{-3} cm^2. Such area has been used to calculate Γ_o before the hybridization process (by taking the value of the adsorbed charge before adding the target DNA, $Q = 2 \times 10^{-8} \pm 0.3$ C) and from that value the amount of DNA on the surface of the electrode was calculated according to:

$$\Gamma_{\text{DNA}} = \Gamma_o \left(\frac{z}{\text{m}}\right) N_a$$

where Γ_{DNA} is the surface density of the immobilized DNA on the surface of the electrode (molecules/cm^2), *m* is the number of bases in the DNA probe, z is the charge of the ruthenium species and N_a is the number of Avogadro. By using such equation, we have calculated a surface density of probe DNA of 5×10^{12} molecules/cm^2. Then we have calculated the amount of DNA that has gone under hybridization with the probe DNA by using the adsorbed charge corresponding to the plateau of Fig. 6.22 ($Q_{ads} = 4.4 \times 10^{-8} \pm 0.2$ C) and subtracting the charge before hybridization. Thus, the number of target DNA strands hybridized with the probe DNA and the hybridization efficiency can be estimated. We have calculated a value 5.1×10^{12} molecules/cm^2. Those values suggest that almost all the probe DNA has undergone hybridization.

6.2.2.3 Electrochemical Impedance Spectroscopy

Impedance measurements have also been performed to follow the hybridization process. Before going in the details of such approach, we would like to stress an important aspect which has to do with the scaling down of the electrode size and its effect on the impedance spectra. Since we are performing impedance spectra in presence of a redox active reporter, $Ru(NH_3)_6^{2+/3+}$, the more appropriate impedance parameter to probe the interfacial changes is the charge transfer resistance (R_{ct}). However and due to its high electron kinetics, the impedance analysis of interfacial processes on metallic macroelectrodes based on the determination of R_{ct} parameter becomes impossible. In such cases the impedance spectra is fully dominated by the Warburg impedance, that is, the electrochemical process is

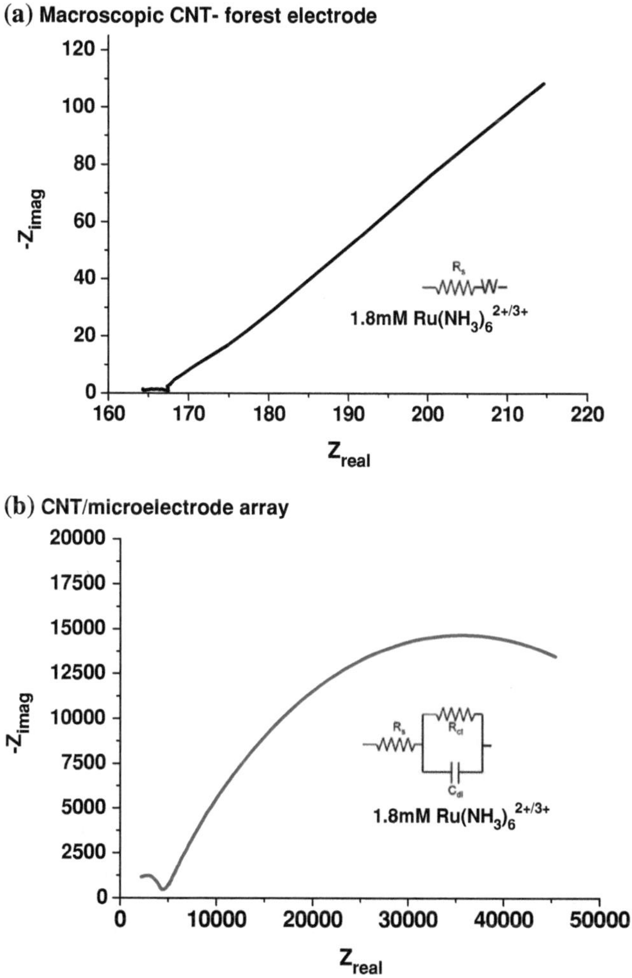

Fig. 6.23 Impedance spectra of **(a)** macroscopic CNT forest electrode and **(b)** CNT microelectrode array in presence of 1.8 $Ru(NH_3)_6^{2+/3+}$. The transition of the impedance spectra from a diffusion-controlled system to a more kinetic-controlled one can be observed as the electrode is miniaturized. The figure also shows the different equivalent circuits which describe the electrochemical response

controlled by diffusion. However, the diffusion control of the electrochemical process can start decreasing when decreasing the size of the electrode [31].

Figure 6.23 shows this transition from diffusion to electron kinetics control due to a decrease in the electrode area and therefore an increment of the system impedance. The Nyquist plot changes from a straight line (Warburg impedance control by using macroscopic forest CNT electrode) to a semicircle (charge kinetics control by using CNT microelectrode array). Such transition can be explained as follows. The charge transfer resistance depends on the exchange current (I_o) at the equilibrium potential of the redox pair ($R_{ct} = RT/nFI_o$) where R is the molar gas

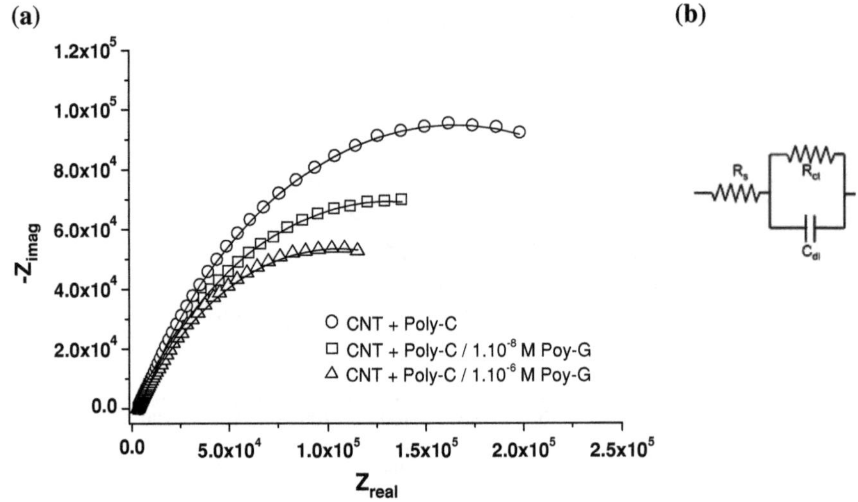

Fig. 6.24 **a** A set of impedance spectra of the probe modified CNT microelectrode performed with 1.8 mM $Ru(NH_3)_6^{3+}$ in PBS with increasing concentrations of target DNA. The electrolyte was deoxygenated with N_2 bubble for 5 min prior to measurements. The impedance was performed with PBS or 1.8 mM $Ru(NH_3)_6^{3+/2+}$ in TRIS at −0.26 V. **b** Equivalent circuit for impedance analysis of hybridization process. R_s solution resistance, R_{ct} charge transfer resistance, *CPE* constant phase element. *Solid lines* represent the fitting of the impedance spectra according to the equivalent circuit

constant, T is temperature and n is the number of electrons involved in the redox reaction. On the other hand I_o depends on the electrode area ($I_o = nFAkC$), where F stands for the Faraday constant, A is the electrode area, k the electron transfer rate constant and C is the concentration of the redox species. In presence of such reversible redox pair, the change of the limiting process is dominated by the electrochemical active area. On macroscopic electrodes, the electrochemical active area is so large that results in a very small charge transfer resistance. Under these conditions, the electrochemical process is limited by the diffusion of the redox species to the electrode. However, when the electrochemical active area is decreased, R_{ct} is increased in such a way that the electrochemical response begins to be dominated by the electron transfer process itself. Such change in the control process at the electrochemical interface has a real advantage in our case. The impedance detection of the hybridization process based on the charge transfer resistance of the ruthenium reporters is now possible if CNT microelectrodes are used. Under these conditions, one can take the R_{ct} parameter as indicative of the hybridization process and develop a practical, reversible and simple protocol based on $Ru(NH_3)_6^{3+}$.

Thus, the hybridization process has been monitored on the CNT microelectrode array. An example of the impedance spectra during the biorecognition events is collected in Fig. 6.24. The parameter R_{ct} was obtained by fitting the impedance spectra to the equivalent circuit depicted in the same figure. The relative error of such parameter with respect to the fitting was found to be about 0.1–0.04 %. Figure 6.24a also includes the fitting curves in solid lines. At this point it is important to mention that the fitting of the results was done with an equivalent circuit that involves constant phase elements (CPE) for the capacitance of the double layer instead of purely capacitive elements. The use of CPE elements are generally preferred since take into account the non-homogeneities of the electrode surface. The impedance of a CPE element is given by $Z_{CPE} = C^{-1}\,(j\omega)^{-n}$, where C is the capacitance and n is the fitting parameter that can vary from 0 to 1. When $n = 1$, the constant phase element approaches the value of a pure capacitance. In our case, n was varying between 0.78 and 0.73.

The quantitative values of R_{ct} have been plotted in Fig. 6.25 as a function of increasing target concentrations. As mentioned before, this parameter provides us information about the easiness for electron transfer at the electrode interface and the trend of this parameter is similar to the findings of the cyclic voltammetry. R_{ct} decreases as the cationic redox marker is accumulated on the interface due to the surface increase of negatively charged phosphates. The electrostatic attraction of the redox marker facilitates the electron exchange as the hybridization process is taken place. The change in resistance is again very sensitive at concentrations below 2.5×10^{-7} and then reaches saturation. Limits of detection were calculated to be around 1.0×10^{-10} M (taken as three time de standard deviation of the blank sample), showing the highly superior sensitivity of this technique.

6.2.3 General Conclusions of Direct (Label-Free) DNA-Detection

The DNA-detection scheme profits from the CNT versatility to develop new miniaturized electrode configurations and also from the CNT suitability to chemical DNA wiring. In fact, vertically aligned CNT electrodes have shown to be a convenient platform for single point linkage of probe DNA and for following up the hybridization process. The direct electrochemical detection of DNA methodologies such as the oxidation of the DNA bases (mainly guanine) or the use of highly positively charged or negative redox reporters as indicators of the hybridization process have been of considerable interest. DNA base-oxidation is attractive since greatly simplifies DNA hybridization assays and offers an instantaneous detection of the duplex, although it is an irreversible process that impedes subsequent use of the electrode for further studies. The positive contribution of CNTs is the enhancement of the oxidation signals due to high surface area effects. We could not observe any decrease in the base oxidation overpotential, a characteristic that would have provided impact to the detection strategy based on DNA oxidation. By evaluating the drawbacks of the direct base oxidation and the limited positive contributions of CNTs, other simple label-free DNA-detection strategies have been tested. The use of electrostatic groove redox binders and electrochemical techniques such as cyclic voltammetry, chronocoulometry and Impedance spectroscopy, have successfully allowed following the hybridization process in a sensitive way. These techniques permit a quantitative estimation of the immobilized DNA probe and its hybridization efficiency. Moreover, we have shown that the scaling down of the CNT electrode size (CNT microelectrodes) results crucial when using impedance measurements under such practical and simple DNA-detection protocol.

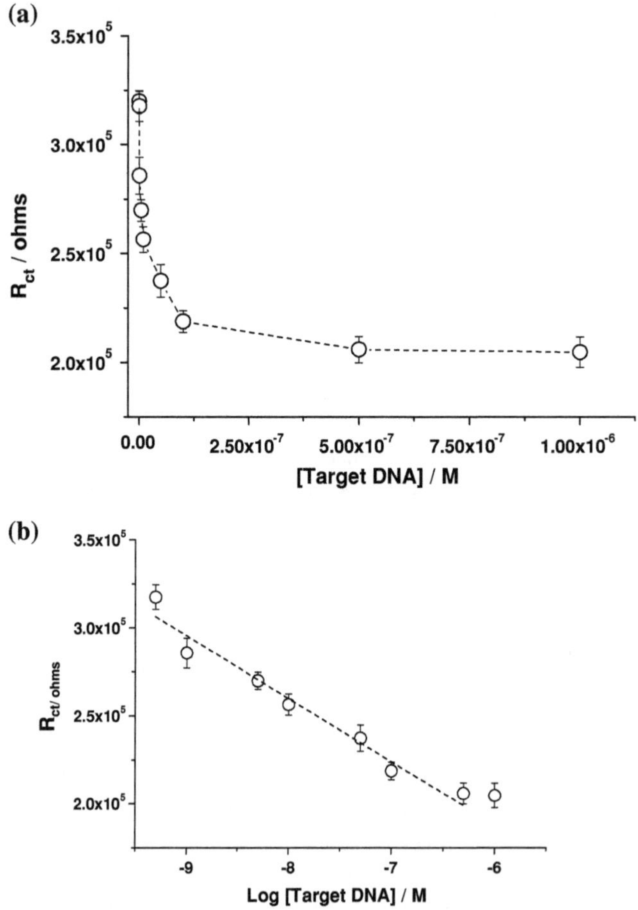

Fig. 6.25 Change transfer resistance extracted from the impedance spectra of the probe modified CNT microelectrode performed with 1.8 mM $Ru(NH_3)_6^{3+}$ in PBS with **a** increasing concentrations of target DNA, **b** logarithm of the increasing concentrations of target. The electrolyte was deoxygenated with N_2 bubble for 5 min prior to measurements. The impedance was performed with PBS or 1.8 mM $Ru(NH_3)_6^{3+/2+}$ in Tris at −0.26 V potential, 10 mV amplitude, from 300 mHz to 100 kHz lower and upper limit. *Error bars* correspond to triplicate measurements

6.3 Aptamers as a Molecular Recognition Elements for Impedimetric Protein Detection

In this section we will continue evaluating the performance of CNT electrodes for monitoring biorecognition events by using vertically aligned CNT microelectrode arrays and CNT modified glassy carbon electrodes. These electrode platforms will be specifically used for following up aptamer-protein detection. For accomplishing that, the electron transfer resistance changes due to protein binding onto

CNT-aptamer modified surface will be measured using electrochemical impedance spectroscopy. For optimizing the functionalization process and the measurement procedure, we have taken the electrode system based on CNTs deposited on glassy carbon electrodes as a reference starting model. This system was also used because of its simplicity and because it was easy-to obtain a new reproducible renewed electrode in few minutes to perform additional experiments during the optimization process.

As stated in Chap. 1, aptamers, artificial nucleic acid with specific binding affinity and selectivity for amino acids, drugs, proteins and other small molecules, have potential applications as a recognition elements in analytical and diagnostic assays [32, 33]. Such species are getting more attention as more robust capture molecules compared to current sensitive antibodies or enzymes, since their special characteristics such as long term storage, thermal stability, etc., can be very useful for realizing easy-to-stock and easy-to use biosensors.

Aptamers which interact specifically with thrombin and lysozyme are among the most studied ones due to the relevancy of both proteins in physiology and pathology. Lysozyme's relatively small size and simplicity makes it an excellent model analyte for novel methods in protein detection; as well as thrombin which is also chosen as a model of target analyte to demonstrate the feasibility of different methodologies.

6.3.1 Functionalization Scheme of the Aptasensor

Below, we show a schematic diagram of the different steps in the development of an electrochemical aptasensor for the detection of thrombin and lysozyme proteins (Fig. 6.26). CV and EIS were mainly used for monitoring each step of the fabrication process in presence of the redox reporter $[Fe(CN)_6]^{3-/4-}$.

Again, and thanks to the experience gained with DNA immobilization techniques, we have used the carbodiimide chemistry to bind aptamers on the CNT surface, which allows, as stated, establishing an amide bond between −COOH functional groups of the CNT and amine terminated oligonucleotide probes. Following this strategy, microelectrodes were carboxyl functionalized by electrochemical pretreatments; while the CNTs used to modify the GCE were oxidized by heat treatment and HNO_3/HCl immersion as discussed in the experimental part.

As previously mentioned in the experimental part, we performed an activation process to the aptamer before anchoring to the nanotube. That process, consisting in a denaturation step, breaks any pre-existent folding structure and confers flexibility to assure proper folding when interacting with specific targets. We will show later on how the activation process can dramatically affect the impedimetric response of the sensor.

In order to verify the amide bond formation between the aptamer and the carboxylic groups of the CNTs, XPS spectra were performed (Fig. 6.27). The XPS data show two different electron binding energies: the first one, around 400.5 eV

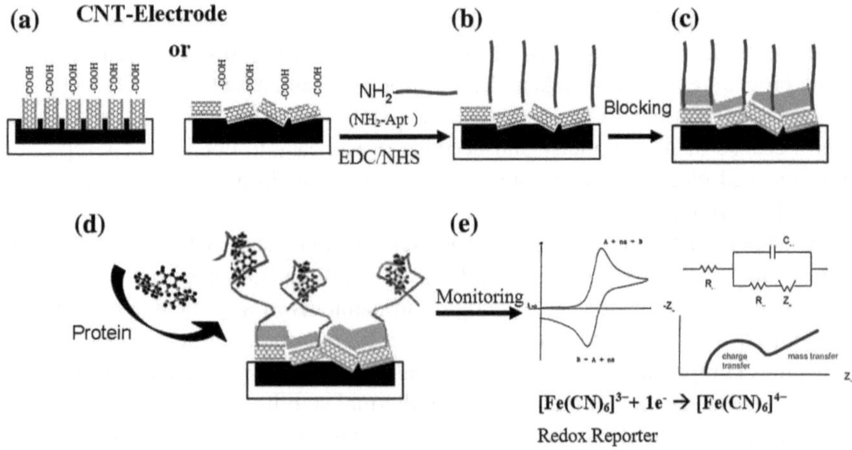

Fig. 6.26 **a** Activation of the CNT electrode surface; **b** binding the recognition element; **c** prevention of non-specific adsorption; **d** target protein; **e** monitoring the sensing event

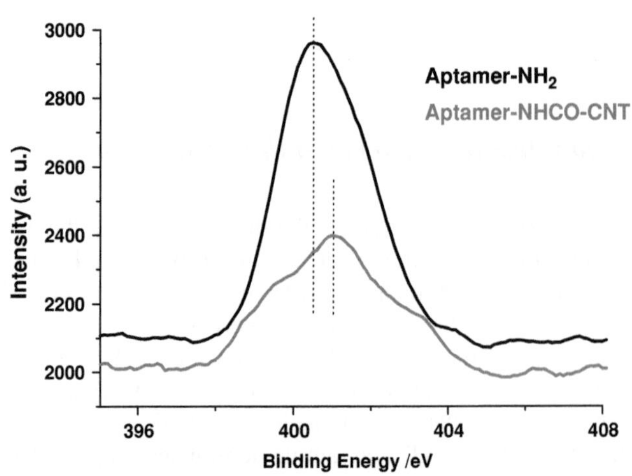

Fig. 6.27 XPS spectra of the aminated aptamer and the amide formation with CNT

(black spectra) comes from a reference sample which only contains the aminated aptamer; in the second one (grey spectra) a shift of around 0.5 eV can be observed with respect to the aminated aptamer which is due to the amide formation between carboxylic moieties of the carbon nanotubes and the aminated aptamer.

These results demonstrate that aptamer probe is successfully fixed on the electrode and that the sensing interface is obtained.

After that step we have addressed the important topic of non-specific adsorption of biomolecules which can bring false-positive results and reduce the accuracy

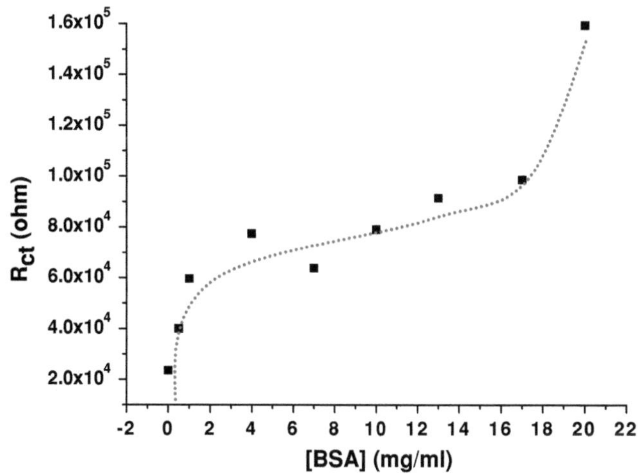

Fig. 6.28 Optimization of blocking reagent concentration. Charge transfer resistance (R_{ct}) variation with BSA concentration in CNT microelectrode array

of the bioanalysis [34]. To achieve specific recognition and decrease non-specific protein adsorption we have tested different blocking agents such as bovine serum albumin (BSA) and poly(ethylene glycol) (PEG). Although of great significance, a thorough and systematic investigation on the blocking agents for avoiding the critical issue of non-specific adsorption has not been covered in this thesis. However, it keeps as an open and very important issue to be strengthened in future studies.

One of the aspects to take into account in order to block a surface and prevent it from non-specific adsorption is the concentration of blocking reagent. One should choose an optimum concentration that could cover the free space that is not occupied by the aptamer but without generating multilayers. Often, molecules do form monolayers following Langmuir theory [35] (monolayer molecular adsorption), or multilayers following BET theory [36] (multilayer adsorption). We have performed a sort of adsorption isotherms of the blocking agent on the CNT surfaces. As an example, Fig. 6.28 shows the adsorption curve of BSA on a CNT microelectrode in terms of the impedance parameter, charge transfer resistance (R_{ct}). In this case, R_{ct} represents the interfacial resistance to electron transfer from the protein modified electrode to the redox couple ferricyanide in solution and it was quantitatively obtained from fitting the impedance data with a Randles equivalent circuit. Different concentrations of albumin were added on the CNT microelectrode array functionalized with a thrombin aptamer and R_{ct} was calculated from the impedance response. Such BSA adsorption curve seems to follow the BET theory. The initial rapid rise corresponds to the formation of the first layer. The second rapid rise at higher concentrations can be ascribed to a second layer of BSA. Thus, such kind of studies allows us to estimate the concentration of blocking

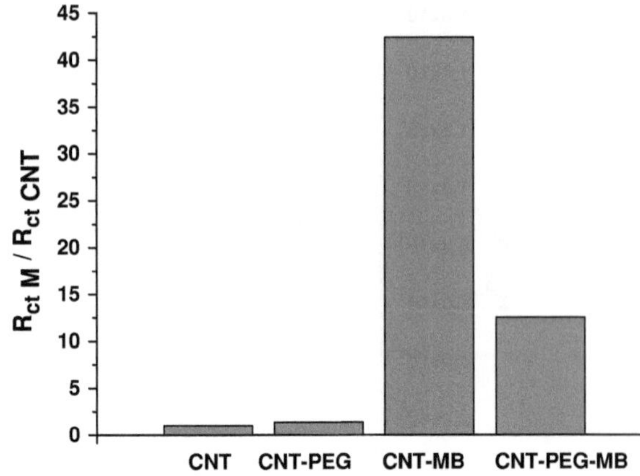

Fig. 6.29 CNT/glassy carbon system. Myoglobin (Mb, at a concentration of 1.0×10^{-9} M) adsorption represented as charge transfer resistance (R_{ct}) when the electrode is treated or not with PEG

reagent necessary to have a covered surface and at the same time to avoid a multilayer adsorption that could have influence in blocking the recognition event. Accordingly, we have chosen a concentration of 1 % (10 mg/ml) for BSA. The same type of measurement was done for a PEG modified electrode and a concentration of 0.04 M was chosen for PEG.

Additionally, we have also followed up the efficiency of the blocking agents to protein adsorption. As an example Fig. 6.29 illustrates the effect of PEG to the adsorption of a model protein such as myoglobin on a CNT modified glassy carbon. The figure shows a normalized charge transfer resistance as a function of different modified CNT electrodes. The normalized R_{ct} has been expressed as the ratio between R_{ct} of the modified CNT/glassy carbon system (R_{ctM}) and the R_{ct} of the unmodified CNT/glassy carbon electrode (R_{ctCNT}). A significant increase of the charge transfer resistance can be observed when myoglobin is added on an unblocked surface (without PEG). However, when the surface is covered by PEG the impedance resistance increases in a small amount by adding the same concentration of myoglobin. Such small increase is due to the fact that the blocking agent can not avoid the 100 % of the non-specific adsorption of proteins.

6.3.2 *Sensor Response to the Aptamer Activation and Non-specific Adsorption*

More insights of the issues discussed above can be collected by performing an analysis of the sensor sensitivity from calibration curves of the aptamer modified CNT system.[2] For instance, we have evaluated in more detail the effect of the

[2] Just for recalling, the sensitivity can be calculated from the slope of the sensor calibration curve.

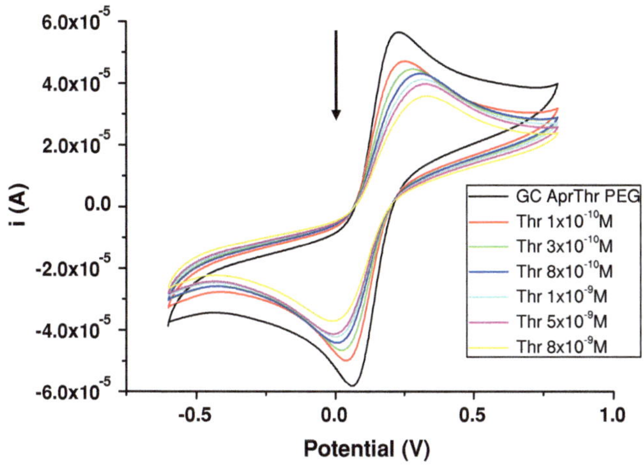

Fig. 6.30 Cyclic voltammograms showing a decrease of peak current, slightly increase of peak separation (increase of R_{ct}) with increasing concentration of thrombin to the system GC/CNT-aptamer thrombin-PEG. Measurements were taken in a 2 mM $K_3Fe(CN)_6/K_4Fe(CN)_6$ solution. Scan rate: 100 mV/s, electrolyte 1× PBS (pH = 7.4)

aptamer activation process and the blocking agent performance for the non-specific adsorption. The calibration curve has been acquired by plotting again a normalized R_{ct} as a function of increasing concentrations of the target protein. In this case, the normalized R_{ct} has been expressed as the ratio $R_{ct}S/R_{ct}P$, where $R_{ct}S$ is the charge transfer resistance of the modified CNT sample with aptamer and blocking agent after binding with its target protein and $R_{ct}P$ is the charge transfer resistance of the sample modified with aptamer probe and blocking agent but before binding with its complementary protein. In general, an increase of charge transfer resistance ratio is observed together with a decrease in peak current in cyclic voltammetry (Fig. 6.30) which is associated with the amount of the selective binding of the protein with the aptamer. The increase of the resistance can be ascribed to the immobilization of the protein at the electrode interface which sterically hinders the electron exchange of the redox ferricyanide couple at the interface. Coulomb repulsive interactions between the negatively charged redox couple and the charged interface could also explain a resistance increase; but in our experimental conditions, working at the isoelectric point of the protein, we considered them to be negligible.

Figure 6.31 shows the response of a CNT/glassy carbon electrode modified with lysozyme aptamer to increasing concentrations of lysozyme with and without the previous aptamer activation process. As can be observed, the activation of the aptamer yields to a dramatic increase of the sensor sensitivity (3.1×10^9 M^{-1}) with respect to the inactivated one (7.3×10^8 M^{-1}). Consequently, such activation step was always performed in all the following measurements.

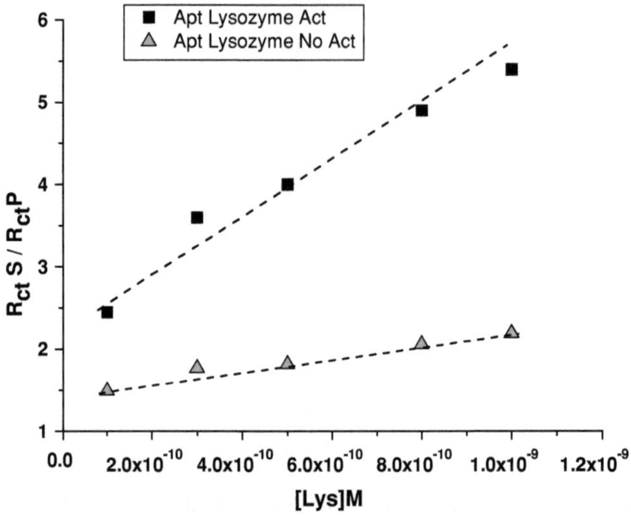

Fig. 6.31 System CNT/glassy carbon with activated aptamer of lysozyme versus non activated aptamer of lysozyme, pH = 11. $R_{ct}S/R_{ct}P$ [charge transfer resistance of the blocked sample ($R_{ct}S$) after binding with the target protein/charge transfer resistance of the blocked sample before binding with the complementary protein ($R_{ct}P$)] as a function of increasing concentration of lysozyme

The blocking agent performance for the non-specific adsorption was then evaluated. Figure 6.32 shows the effect of the blocking agent on the aptasensor made from either modified CNT/glassy carbon system or modified CNT microelectrode array. We have used PEG and BSA as candidates for the non-specific protein adsorption and we have observed that the sensor sensitivity is greater when PEG is used as blocking agent than when using BSA. For instance in the modified glassy carbon system a sensitivity of 3.11×10^9 M^{-1} was obtained when using PEG as compared with 2.05×10^8 M^{-1} when using BSA. Similar trends were observed in the modified CNT microelectrode systems. We obtained a sensitivity of 2.9×10^9 M^{-1} using PEG versus 9.3×10^8 M^{-1} using BSA. Such results encourage us to choose poly(ethyleneglycol) as blocking agent for non-specific protein adsorption. It is quite probable that the big size of the BSA protein can mask recognition signals by sterically hindering the aptamer activity.[3]

6.3.3 *Aptasensor Response to Ionic Strength and pH*

After evaluating the aptamer activation step and the non-specific adsorption of the blocking agent, we have addressed the effect of ionic strength and pH on the sensor sensitivity and selectivity.

[3] BSA has a large molecular weight of 60 kDa; twice the molecular weight of target proteins as for instance thrombin. BSA could bury the signal of the low molecular weight aptamer.

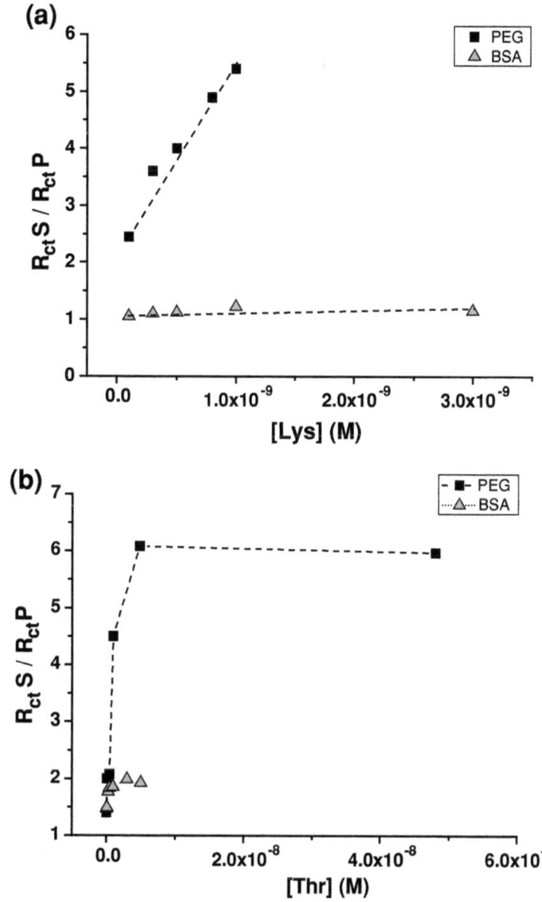

Fig. 6.32 **a** GG/CNT—AptLys-PEG versus AptLys-BSA. Lys, pH = 11 with increasing concentration of lysozyme. **b** CNT microelectrode array—AptThr-PEG versus AptThr-BSA, pH = 7 with increasing concentration of thrombin

With respect to the ionic strength, the increase of NaCl concentration in the electrolyte decreases the sensor sensitivity. That effect could be connected with either the shielding of the negative charges at DNA aptamer as well as at protein surface or with changes of the conformation of the binding site of the aptamer. Some studies claim that at higher concentration of Na^+ the proteins could even aggregate, which may cause its lower affinity to the aptamer. Nevertheless, and in connection with conformation changes of the aptamer induced by metallic ions, previous studies found that in the case of thrombin aptamers the presence of K^+ results beneficial for the stabilization of the appropriate protein binding conformation of the aptamer. Accordingly, we have found an optimized ionic strength in our PBS buffer by working at lower concentration of NaCl (50 mM) and with 8 mM of KCl.

With respect to the pH issue, we have first evaluated the aptasensor response for lysozyme protein and for non-specific proteins such as cytochrome C at

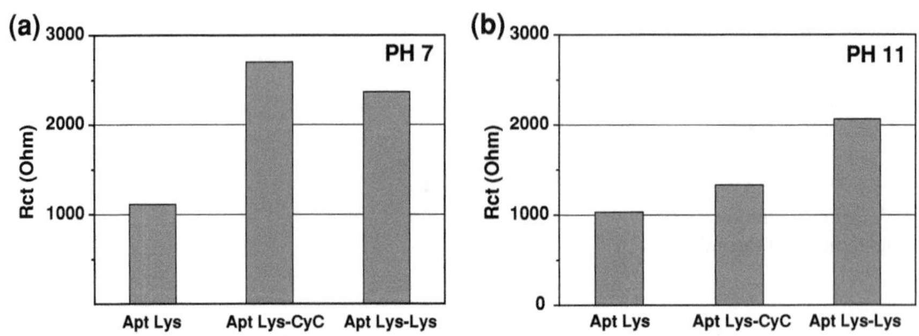

Fig. 6.33 Histograms showing pH independence to protein-aptamer affinity in GC system. R_{ct} obtained from data fitting of EIS spectra with the Randles equivalent circuit

physiological pH in the glassy carbon system. Lyzosyme (Lys) and cytochrome C (cyC) are similar molecular weight proteins, very basic with similar isoelectric point: Lys around 11 and cyC around 10.7. Figure 6.33a shows the impedance response at pH = 7. Surprisingly, we found large interaction between the lysozyme aptamer with the non-specific target cyC as compared to its own specific protein. Such positive sensor response with cyC could be due to a larger extent of non-specific protein adsorption under such experimental conditions. The non-specific adsorption can be mediated by different kind of interactions (hydrophobic, Van der Waals, electrostatic force, etc.) which can be increased depending on the pH conditions. At neutral pH, both proteins are positively charged. Only taking into account the charge distribution in amino acid side chains, lysozyme has 17 positive charges and 9 negative charges, resulting in a net +8 charge [37] whereas cyC has 19 positively charged lysine residues, plus two arginines also positively charged, but only 12 acidic residues (aspartic or glutamic acids) resulting in an net +9 charge [38]. This may cause an electrostatic change of affinity of the interfering cyC towards the negatively charged AptLys[4] at pH = 7 which results in a more specific adsorption. We also found that such effect is decreased when working at the isoelectric point of the protein as depicted in Fig. 6.33b. In such figure, one can see a higher affinity of lysozyme aptamer for its specific protein as compared to the non-specific one. Such results also show that the pH is a very important parameter to tune the selectivity of the aptamer–protein interactions. According to these results and trying to avoid electrostatic mediated adsorption, we have decided to work at pH close to the isoelectric point of the target protein.

Similar results have been found when working with thrombin aptamer modified sensors. It was observed that maximal thrombin response took place for pH 7.5 (close to its isoelectric point), while for lower and higher pH the sensitivity was lower. Figure 6.34 shows the impedimetric response of CNT/glassy carbon

[4] Coulombic attraction between the phosphate containing backbone of nucleic acids and the positively charged protein could increase association facilitating the immobilization of protein at the electrode interface.

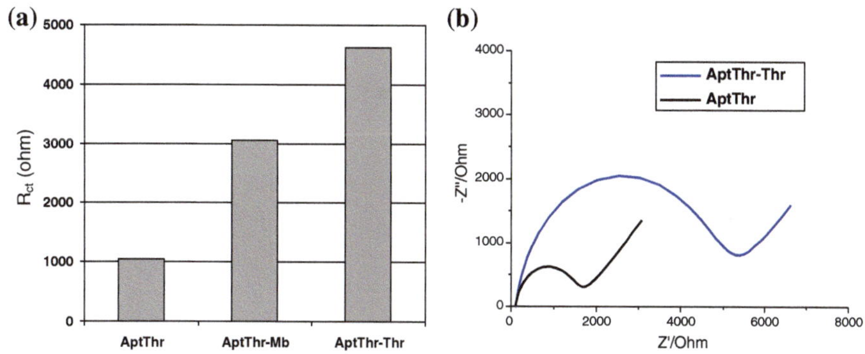

Fig. 6.34 **a** Histograms to validate the selectivity of the system (AptThr is the aptamer of thrombin, Mb is myoglobin and Thr is thrombin). **b** Nyquist plot: one can observe the increase of the charge transfer resistance (increase of the *semicircle* diameter of plot) when interacting with its specific protein

electrode functionalized with the aptamer of thrombin at pH = 7 in presence of its specific complementary protein (thrombin) and non-complementary one (myoglobin), both of them at a concentration of 1.0×10^{-9}M. The figure also depicts the Nyquist impedance spectra showing the electrode response of the aptamer modified electrode and when interacting with its target protein.

It can be observed that the aptasensor impedance response increases when exposing to the complementary protein as compared to the non-complementary one. However, the sensor also exhibits a response to the non-specific protein (Mb) which can be rooted again from some non-specific adsorption.

Although the total elimination of non-specific adsorption is impossible, we have tried in these studies to minimize such effect by choosing appropriate blocking agents, appropriate concentrations and pH conditions. However and in light of our results, the optimization of this issue needs more experimental research investment and even more in porous electrodes such as the ones we are dealing with. Porosity can bring more defective coverage of the blocking agent and also the porous nature can induce, at certain extent, protein inclusion or entrapment favouring the non-specific interactions.

6.3.4 Electroanalytical Parameters of the Aptasensor

Finally and after evaluating the response of the aptasensor to different experimental conditions, we carried out a closer inspection of the aptasensor by performing calibration curves at the more optimum conditions. Figure 6.35 shows the sensor characteristics of CNT/glassy carbon modified with thrombin aptamer, blocked with PEG and at pH = 7.4. Such modified electrode has been subjected to different concentrations of thrombin and to a non-specific target such as elastase.

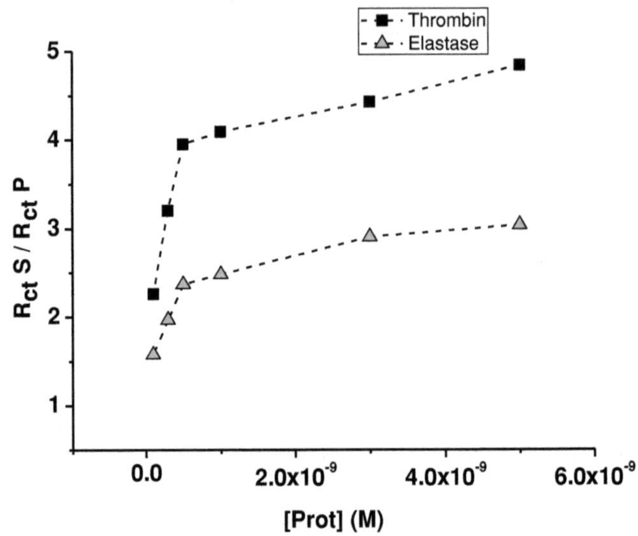

Fig. 6.35 Selectivity of the system GC/CNT-AptThr-PEG with thrombin versus elastase at pH = 7.4

The system exhibits remarkable sensitivity at the low protein concentration range (5.0×10^{-11} to 1.0×10^{-9} M). It was found a sensitivity of 5.5×10^{9} M^{-1} with the target protein thrombin and limits of detection in the order of picomolar, 37.5×10^{-12} M^{-1}. The limit of detection was calculated by considering the expression $\frac{1}{R_{ct}S} \cdot \frac{3\sigma_{blank}}{S}$, where σ_{blank} is the standard deviation in blank solution and S is the sensitivity calculated from the slope of the sensor response $R_{ct}S/R_{ct}P$ versus protein concentration.

The sensitivity to elastase interaction is half to the one found with thrombin. That evidences higher interaction with the specific protein though non-specific adsorption can not be eliminated.

We have also evaluated the sensor response of the CNT/glassy carbon electrode modified with lysozyme aptamer as a function of increasing concentrations of lysozyme and at pH = 11. Figure 6.36 shows the calibration curve together with the impedance Nyquist spectra. It can be again observed that the electrode system is very sensitive to concentrations below 1 1.0×10^{-9} M. Above such values the electrode reaches saturation. Similar sensitivities (3.15×10^{9} M^{-1}) and limits of detections (around 53 pM) were found as compared to the thrombin aptamer modified sensors.

Similar analyses were performed on aptamer modified microelectrodes. As an example, Fig. 6.37 shows a calibration curve of a thrombin aptamer modified CNT microelectrode as a function of increasing concentrations of thrombin. The results show similar results as compared with the ones described above: remarkable sensitivity at concentrations below 1.0×10^{-9} M, sensitivities of 3.5×10^{-9} M^{-1} and limits of detections in the picomolar range (68 pM).

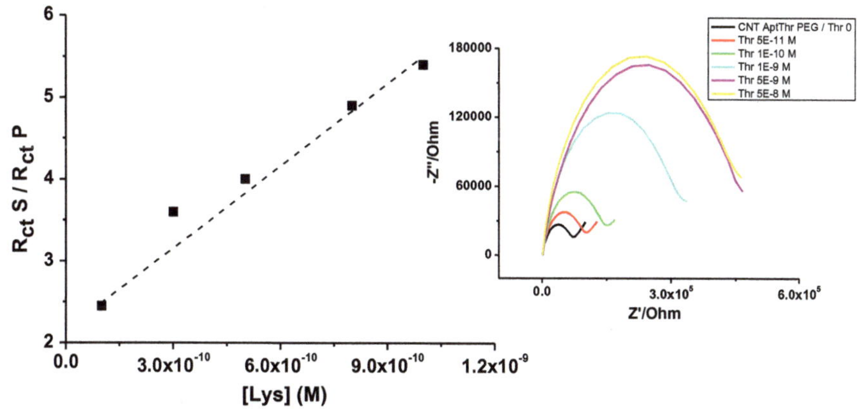

Fig. 6.36 Calibration curves to determine the sensor response of the GCE/CNT-AptLys system. Representation of $R_{ct}S/R_{ct}P$ versus concentration of protein (*left*); representation of Nyquist plot from EIS measurements (*right*) (pH = 11)

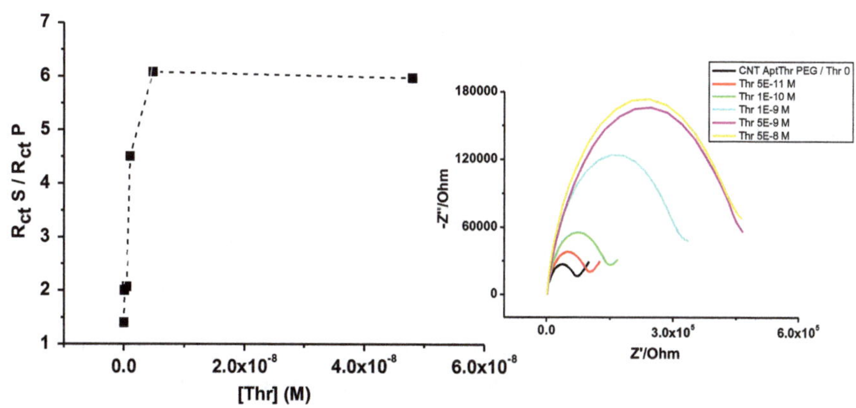

Fig. 6.37 Calibration curves to determine the sensor response of the CNT microelectrode-Apt-Thr system. Representation of $R_{ct}S/R_{ct}P$ versus concentration of protein (*left*); representation of Nyquist plot from EIS measurements (*right*) (pH = 7.4)

6.3.5 General Conclusions of the Impedimetric Protein Detection with CNT-Based Aptasensors

In conclusion we have observed that the aptasensor impedimetric response is very dependent on the previous treatment of the aptamer, the blocking agent for non-specific adsorption and the ionic strength and pH of the buffer solution. We have presented here our first preliminary efforts for tuning all these aspects to achieve appropriate sensor response. However, there are still many open issues that need further optimization for a better sensor performance. One of them is the topic of the non-specific adsorption minimization which can be improved by testing other blocking agents or addressing CNT platforms with lower degree of porosity. A lower degree of porosity could even decrease the limits of detection. In spite of the present limitations, both CNT sensor platforms have provided good limits of detection, but perhaps the microelectrode CNT configuration represents a more promising candidate since impedance signal is amplified with miniaturization and at the same time is a more tuneable system for controlling porosity or for generating CNT ultramicroelectrode arrays of better analytical performance as discussed in the following section.

6.4 UMAs a Step Forward for Ultrasensitive Detection

In previous sections, we have dealt with forest CNT microelectrode arrays. Although they have shown some advantages for biosensing purposes, such as in the impedimetric detection, still behave like macroelectrodes. However, their electrochemical performance can be increased by increasing miniaturization and thus approaching the ultramicroelectrode regime. Therefore in this section we will introduce the electrochemical response of ultramicroelectrode devices whose fabrication steps were detailed in the experimental part. Although this topic has been incorporated in the section on CNT platforms for biosensing, biorecognition events with such CNT electrodes have not been achieved yet. However, the electrochemical aspects of UMAs will be addressed here as a natural continuation of the CNT platforms discussed before and focusing on the potential perspectives for future studies in the electrochemical biosensing.

In Chap. 1, we have pointed out the benefits of using ultramicroelectrode arrays for electroanalytical purposes. Just for recalling, they promote high mass transfer rate of the analyte in the solution to the electrochemical interface provided by a fast radial diffusion. Due to that, the electrode quickly reaches a steady state current which is directly proportional to the analyte concentration. That makes the electrode exhibit a very fast response, very high signal/noise ratio, and consequently very low limits of detection, being all these properties highly desired in amperometric sensors. In Chap. 1 we also indicated the difficulties in manufacturing ultramicroelectrode arrays and in this thesis we proposed a methodology for fabricating ultramicroelectrode arrays from CNTs based on the opal inversion technique, which was thoroughly explained in the experimental part. Carbon nanotubes represent a feasible alternative to the (ultra)microelectrode arrays made from conventional metals and also this novel fabrication strategy demonstrates the versatility of this material for generating new electrode configurations and for tuning their geometrical disposition to desire.

The mentioned technique has allowed us to accomplish vertically oriented CNT arrays (passivated at the walls) with controllable distance separation among the CNT electroactive to avoid the diffusion layer overlapping. Bare CNT arrays can be degraded with multiuse due to the fact that adhesion to the support material

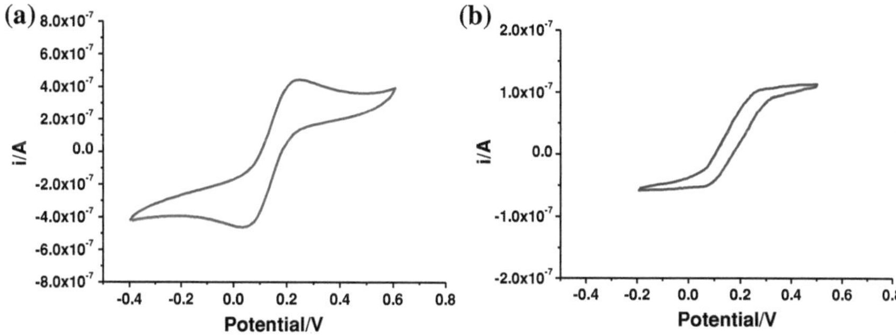

Fig. 6.38 Cyclic voltammetry curves for a high density forest-like microelectrode array (**a**) and ultramicroelectrode array with a low-density patterned VACNTs by the opal inversion technique (**b**). Curves were taken in 4.3 mM $K_4Fe(CN)_6$ in a 1.0 M KCl at 20 mV/s scan rate. In case (a) the electrochemical active spots overlap generating a planar diffusion of the electroactive species and the CV response is in a peak-shaped profile. In case (b) the electrochemical active spots of the electrode are small and sufficiently far away that the diffusion profiles are hemispherical and very dependent on the perimeter of the active surface. That brings about an enhanced and fast mass transport of the electroactive species and consequently fast response and higher faradaic signal with respect to the background noise

becomes sometimes weak. However, a proper passivation in between the electroactive nanoelectrode centres can provide robustness to the CNT arrays. Such passivation can also increase signal/noise ratio by leaving only the terminal edges of the nanotubes electrochemically active, eliminating any wall contribution (the less electroactive part of the CNTs). Thus, the more electrochemical reactive sites of the tube will be the ones facing the analyte solution and will become the chemical active centres for a more controllable (bio)functionalization. Proper passivation between the nanoelectrodes can also reduce the porosity degree. Porous electrodes can introduce compromise situations for electrochemical biosensors. On one side, they can induce electrochemical signal amplification due to higher surface area effects but on the other side, they can contribute with capacitive noise or be detrimental when trying to avoid non-specific adsorption for biorecognition events such as the examples discussed in the previous sections.

Figure 6.38 shows comparatively cyclic voltammograms for a forest CNT microelectrode array and for CNT ultramicroelectrode arrays based on the opal inversion technique in presence of $Fe(CN)_6^{3-/4-}$. In the first case (a), the size of the microelectrode is not sufficiently small as compared to the distances between neighbouring microelectrodes which generates overlapping of their diffusion layers. Thus, such CNT microelectrodes behave similar to a macroelectrode with anodic and cathodic peak currents and high charging/discharging (background) currents, typical of a planar diffusion of the electroactive species. In contrast, CNT ultramicroelectrode arrays show a more sigmoidal CV curve with a steady state current dominated by radial diffusion. Their sizes are smaller as compared to their individual nanoelectrode interdistance which assures negligible overlapping of the diffusion layers from neighbouring electrodes and each CNT spot truly behaves as a single nanoelectrode.

6.4.1 General Conclusions of CNT Ultramicroelectrode Arrays for Ultrasensitive Detection

The results presented here are very preliminary and their quality needs to be considerably increased. Nevertheless, we wanted to establish the first foundations in the development of such electrode systems. Therefore, this issue provides an open research line which needs to be improved and exploited given the promising characteristics of these systems in biosensor technology. Such passivated and upright CNT nanoelectrode arrays with the more electrochemical reactive sites facing towards the analytes can greatly improve the sensitivity and limits of detection up to the range of few molecules. Additionally, they can act as molecular wires where electroactive biomolecules can be plugged at their ends facilitating the direct electron transfer and with the additional feature that the porosity degree or surface area effects can be more controllable for improving selectivity in biorecognition processes.

References

1. Yu X, Chattopadhyay D, Galeska I, Papadimitrakopoulos F, Rusling JF (2003) Electrochem Commun 5:408–411
2. Zhao L, Liu H, Hu N (2006) J Colloid Interface Sci 296:204–211
3. Wang J, Li M, Shi Z, Li N, Gu Z (2002) Anal Chem 74:1993–1997
4. Zhang L, Zhao G-C, Wei X-W, Yang Z-S (2005) Electroanalysis 17:630–634
5. Zhao G-C, Zhang L, Wei X-W, Yang Z-S (2003) Electrochem Commun 5:825–829
6. Zhao G-C, Yin Z–Z, Zhang L, Wei X-W (2005) Electrochem Commun 7:256–260
7. Nassar A-EF, Zhang Z, Hu N, Rusling JF, Kumosinski TF (1997) J Phys Chem B 101:2224–2231
8. Zhang H, Hu N (2007) Biosens Bioelectron 23:393–399
9. Kumar SA, Chen S-M (2007) Talanta 72:831–838
10. Yan F, Ruan C, Chen X, Deng J, Kong J (1999) Fresenius J Anal Chem 363:83–87
11. McCreery RL (1991) In: Bard AJ (ed) Electroanalytical chemistry, vol 17. Dekker, New York
12. Laviron E (1979) J Electroanal Chem Interfacial Electrochem 101:19–28
13. Li Y, Lin X, Jiang C (2006) Electroanalysis 18:2085–2091
14. Zhang L, Zhao G-C, Wei X-W, Yang Z-S (2004) Chem Lett 33:86–87
15. Zhao L, Liu H, Hu N (2006) Anal Bioanal Chem 384:414–422
16. Nakajima R, Yamazaki I (1987) J Biol Chem 262:2576–2581
17. Onuoha AC, Zu X, Rusling JF (1997) J Am Chem Soc 119:3979–3986
18. Zhang Z, Chouchane S, Magliozzo RS, Rusling JF (2001) Anal Chem 74:163–170
19. Fan C, Plaxco KW, Heeger AJ (2003) Proc Natl Acad Sci 100:9134–9137
20. Pividori MI, Merkoçi A, Alegret S (2003) Biosens Bioelectron 19:473–484
21. He P, Xu Y, Fang Y (2006) Microchim Acta 152:175–186

22. He P, Dai L (2004) Aligned carbon nanotube-DNA electrochemical sensors. Chem Commun 3:348–349
23. Bonanni A, del Valle M (2010) Anal Chim Acta 678:7–17
24. Bonanni A, Esplandiu MJ, del Valle M (2008) Electrochim Acta 53:4022–4029
25. Wang J, Li M, Shi Z, Li N, Gu Z (2004) Electroanalysis 16:140–144
26. Wang J, Kawde A-N, Musameh M (2003) Analyst 128:912–916
27. Drummond TG, Hill MG, Barton JK (2003) Nat Biotech 21:1192–1199
28. Gorodetsky AA, Buzzeo MC, Barton JK (2008) Bioconjug Chem 19:2285–2296
29. Steel AB, Herne TM, Tarlov MJ (1998) Anal Chem 70:4670–4677
30. Andreu A, Merkert JW, Lecaros LA, Broglin BL, Brazell JT, El-Kouedi M (2006) Sens Actuators B Chem 114:1116–1120
31. Siddiqui S, Arumugam PU, Chen H, Li J, Meyyappan M (2010) ACS Nano 4:955–961
32. Hermann T, Patel DJ (2000) Science 287:820–825
33. Kawakami J, Imanaka H, Yokota Y, Sugimoto N (2000) J Inorg Biochem 82:197–206
34. Wang Y, El-Boubbou K, Kouyoumdjian H, Sun B, Huang X, Zeng X (2009) Langmuir 26:4119–4125
35. Langmuir I (1916) J Am Chem Soc 38:2221–2295
36. Brunauer S, Emmett PH, Teller E (1938) J Am Chem Soc 60:309–319
37. Blake CCF, Koenig DF, Mair GA, North ACT, Phillips DC, Sarma VR (1965) Nature 206:757–761
38. Barlow GH, Margoliash E (1966) J Biol Chem 241:1473–1477

Chapter 7
Results and Discussion: Electronic Response of Carbon Nanotube Field-Effect Transistors to Biorecognition Processes

In Chap. 1, we have detailed the potentialities of the CNT-FET configuration for label-free transduction of an analyte/receptor interaction. As mentioned, such interaction can be transduced into a readable electrical signal by following up the conductance changes of a semiconducting CNT.

In this section we will discuss different issues related with diverse aspects of the CNT-FET layout such as the use of liquid gating and the optimization of the device by designing a passivation protocol. After that, we will discuss the application of the device to monitor in real-time the adsorption of a protein on the CNT walls and then we will finally evaluate the performance of the CNT-FET as aptasensor by monitoring the biorecognition event between an aptamer with its specific target protein.

7.1 Electronic Detection in a Field-Effect Transistor Configuration in Aqueous Environment

We have also mentioned that liquid-gated FETs are preferable for biosensing applications as compared to dry measurements with a back gate electrode. The aqueous medium is a suitable environment for a proper biomolecule activity. Moreover the liquid-gated FETs allow real-time detection of biorecognition processes. Just for recalling, the liquid gating voltage is applied between the CNT and a reference electrode immersed in the solution. Such electrode controls the electrostatic potential of the solution which strongly affects the conductivity of the nanomaterial. The gate voltage mainly drops over the thin double layer formed at the CNT,[1] so that the effective gate voltage is indeed the applied voltage. Because the

[1] As said above in the liquid-gated CNT-FET a voltage is applied between the CNT and a reference electrode immersed in the solution. Owing to electrostatics interactions, a layer of counter ions is formed close to the CNT and acts as a dielectric layer (double layer).

M. Pacios Pujadó, *Carbon Nanotubes as Platforms for Biosensors with Electrochemical and Electronic Transduction*, Springer Theses, DOI: 10.1007/978-3-642-31421-6_7,

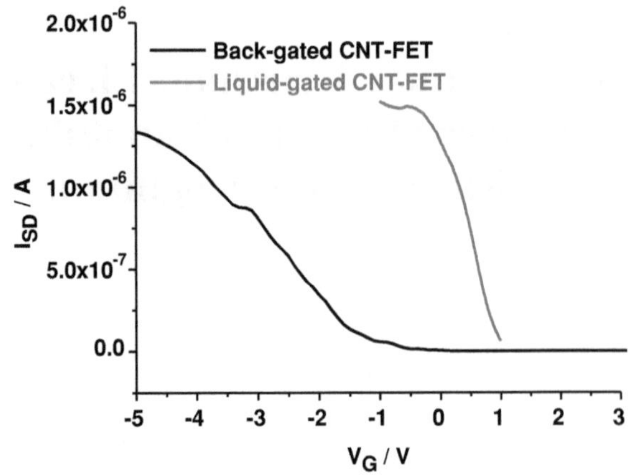

Fig. 7.1 Comparison between a liquid-gated CNT-FET (*grey*) and a back-gated CNT-FET in dry conditions (*black*). The difference in gate coupling strength can be estimated by comparing the voltage axis scaling when changing from a back-gated to a liquid-gated layout

double layer is only a few angstroms thick, much smaller than the thickness of SiO_2 which is widely used in the back gate, a better gate coupling is achieved and consequently the modulation in the conductance of the tube is more effective.[2, 3] Thus, the liquid-gate can become over two orders of magnitude more efficient than the back gate. In Fig. 7.1, a comparison between the conductance response of a CNT-FET with a back-gated electrode in dry conditions and with liquid-gate electrode is exhibited. It can be observed that the V_G scale for the liquid-gate is significantly narrower than that of the bottom-gate arrangements. The large difference in voltage range between the liquid-gated and the back-gated configuration strikingly illustrates the difference in gating efficiency due to the capacitive differences in

[2] The differences in the thickness of the double layer (in the liquid-gated CNT-FET) and the oxide thickness (in the back-gated CNT-FET) have effects in their capacitance values. The lower the thickness of the dielectric the higher the capacitance (the capacitance, C, is inversely proportional to the separation between conducting sheets). That parameter affects in the accumulation of charge (Q) at the capacitor and on the voltage (V) that is needed for accumulating that charge ($Q = CV$). If one wants to accumulate a certain amount of charge it would need lower voltages if the capacitance is large. In the case of a FET operation, higher capacitances can induce more effective charge accumulation and hence more modulation in the CNT conductance. All these effects make that the voltage that is needed to accumulate certain charge and modulate the CNT conductance will be smaller in the case that higher capacitances (smaller dielectric thickness) are involved. In that case it is said that the gate coupling is more efficient and large Fermi level shifts can be achieved.

[3] The total capacitance (C_t) in a CNT-FET system is composed of the gate capacitance (C_d, capacitance of the dielectric, the double layer capacitance in wet conditions) and the quantum capacitance (C_q) of the tube. The quantum capacitance comes from the Pauli Exclusion Principle, which forces electrons to occupy higher energy levels once the lower ones are occupied, leading to an extra increase in the electrical potential. These two capacitances are in series, $1/C_t = 1/C_d + 1/C_q$. It may happen that under certain circumstances in the liquid-gated CNT-FET, C_d can be large enough that C_q can become the dominant term in C_t and the conductance of the tube does not depend significantly on C_d and thus on the liquid composition.

such configurations. The enhanced gate coupling between the liquid and the CNT and the high mobility under liquid environment generate high transconductances which is a crucial factor for determining the ultimate sensitivity.

7.1.1 Passivation Protocol of the CNT-FET Device

In previous sections, we have also discussed that CNT-FETs operated in liquid medium need to be passivated in order to get a good performance. A poor passivation of the device can result in leakage currents or undesired electrochemical currents which can smear the FET response or destroy the electrode contacts by electrochemical oxidation. Accordingly different passivating protocols by using electrochemical techniques were first tested on a model system based on a gold electrode. We have used as passivating layers self-assembled monolayer based on octanethiols, standard polymethyl methacrylate (PMMA) with a molecular weight of 9,50,000 and a copolymer of PMMA with a certain percentage of methacrylic acid (8.5 %). For a fast overview of the blocking properties of such passivating layers, we have deposited them on the gold electrode and used cyclic voltammetry and impedance spectroscopy for the electrode characterization. Figure 7.2 shows the electrochemical data to validate the passivating performance. The figure depicts the cyclic voltammogram of a bare gold electrode and a gold electrode passivated with a self-assembled thiol layer and with 200 nm either of standard PMMA or PMMA copolymer. The experiment was performed in presence of 10 mM of $Fe(CN)_6{}^{3-/4-}$ at a scan rate of 0.1 V/s. When the electrochemical study is performed on bare gold, two peaks are observed related with the oxidation/reduction of the redox couple at the gold surface. When the gold electrode is covered with the insulating layers, the current decreases dramatically whereas the redox peaks disappear as a consequence of the electron transfer blocking provided by passivating layers.

From the voltammograms it can be observed that the insulating layers exhibit very good blocking properties. However, it is also important to evaluate the stability of these layers when polarizing at different voltages, especially at such voltage values at which the conductance of the FET device can be collected under liquid environment (V_G window between ± 1.5V). For instance, Fig. 7.3 shows the impedance behaviour of the passivating layers after polarizing the same gold electrode of Fig. 7.2 at 1.5 V for some seconds. The charge transfer resistance of the redox pair can be taken as a parameter to have insights of the layer stability. As mentioned in previous sections, such parameter is qualitatively related with the diameter of the semicircle of the impedance spectra. It can be observed that after polarization the quality of the passivating layer of the thiol is very poor (due to thiol desorption upon electrochemical oxidation at such voltages) but the PMMA layer provides a high electron transfer resistance keeping its integrity after the polarization. The PMMA copolymer exhibits a lower electron transfer resistance as compared to the standard PMMA. Accordingly we have

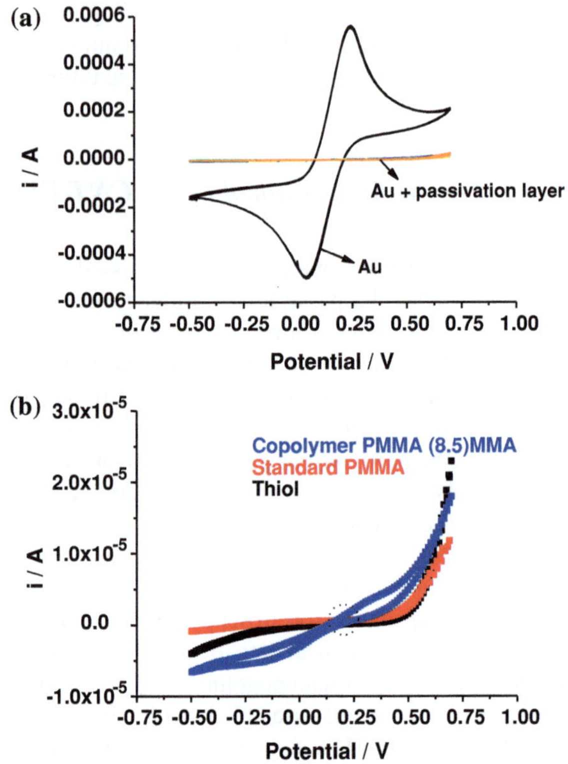

Fig. 7.2 **a** Cyclic voltammetry for a bare gold electrode and with a passivating layer; **b** zoom of the voltammograms of the passivating layers

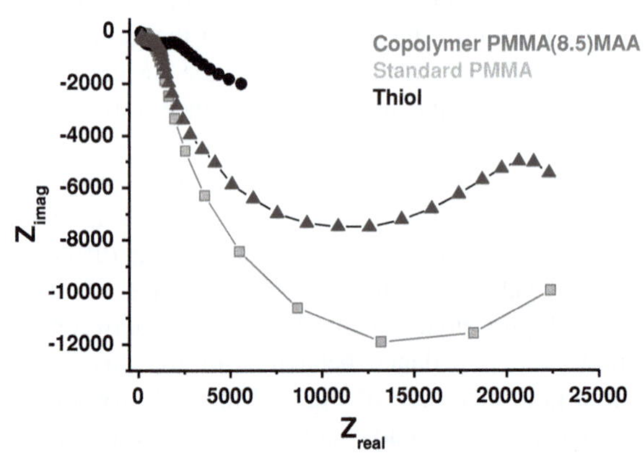

Fig. 7.3 Impedance spectra for a self-assembled thiol monolayer, a standard PMMA layer and a copolymer PMMA layer in presence of $Fe(CN)_6^{3-/4-}$ and after polarizing at 1.5 V

chosen the 200 nm standard PMMA layer as the more optimum passivating layer for the CNT-FET device.

In the experimental section we discussed the general structure of the CNT-FET for liquid gating. As mentioned there, the chip comprises a complex structure

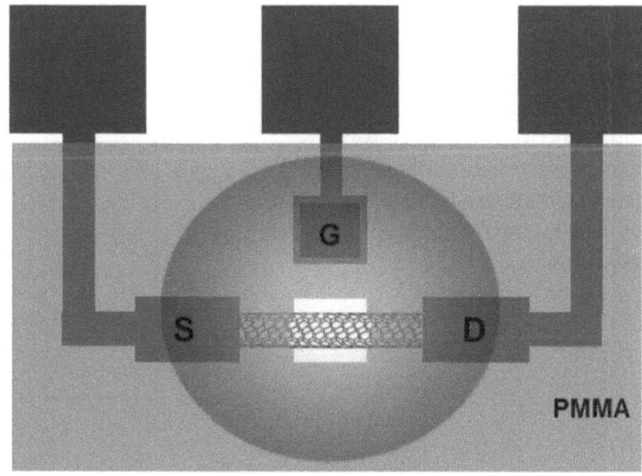

Fig. 7.4 Simplified scheme of the CNT-FET device passivated with PMMA. The passivated layer has been partially removed in the CNT, in the small pad of the liquid-gate electrode (*G*) and on the big contacts of the source (*S*), drain (*D*) and liquid-gate electrodes. The *blue circle* represents an aqueous drop

which has been designed for high production of CNT-FETs (one chip can contain around 100 FET devices). They have been also designed with the possibility of operating under back gate and liquid-gate control. For the latter purpose, the device integrates a gold electrode to control the electrostatic potential of the solution (the electrode for liquid polarization acting as a pseudo-reference electrode) and an appropriate PMMA passivation was performed on the chip. One advantage of having the liquid-gate electrode integrated into the FET device is that it is easy to miniaturize with standard lithographic techniques and that the position of the electrode is always fixed. The disadvantage is that it is not a real reference electrode and under certain conditions can induce an unstable gate potential. Under such situations, the applied gate voltage can not only drop in the tube–liquid interface but also in the gold electrode/liquid interface due to interactions of gold with species in the solution. In the next section we can see that such effect is not so important under the conditions that we have used for our experiments.

A simplified scheme of one CNT-FET device is shown in Fig. 7.4. The sensor is completely passivated by PMMA except for the channel of the CNT-FET, for part of the liquid polarization electrode and for the other contact pads used for the measurements which are not in contact with the liquid.

At this point it is important to state that the FET response obtained in Fig. 7.1 has been acquired under such passivation conditions. We have also performed additional measurements in order to validate the passivation methodology on the CNT-FET itself. Figure 7.5 shows the leakage current (gate current as a function of the gate voltage) due to electrochemical effects with and without passivation. It can be observed how the leakage current decreases when the passivation procedure is carried out. A very low leakage current is a critical parameter for the electrochemical sensing operation of the CNT-FETs since it will determine the sensitivity and stability of the sensor device.

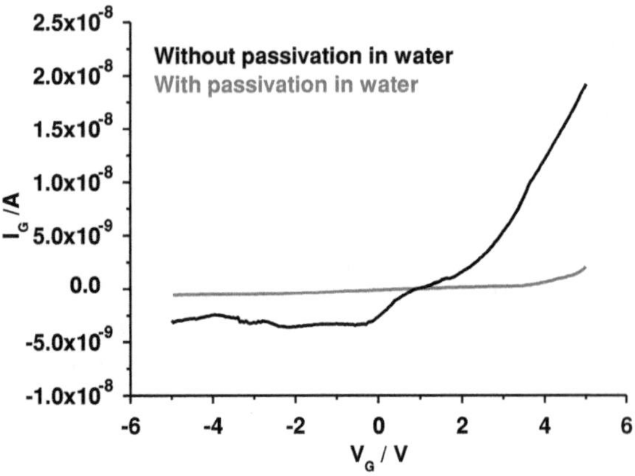

Fig. 7.5 Gate leakage current of a passivated and a non-passivated CNT-FET in aqueous solution. The current between the CNT-Au contact electrodes and the gate electrode can be reduced below pA levels with the passivation

7.2 CNT-FET Layout for Monitoring Protein Adsorption

In order to test the CNT-FET device, we have followed up the changes of CNT conductance upon protein adsorption on the CNT walls under liquid gating. The protein was myoglobin (Mb) which can adsorb non-specifically on CNT walls mainly by hydrophobic forces. Under this context, CNT-FETs can be used as an electronic analytical tool to detect and monitor protein adsorption with high sensitivity.

Accordingly to the discussion detailed in Chap. 1, the mechanisms behind the conductance changes in the CNT-FET characteristics can be rooted from different sources. Changes in the FET conductance can be related to the surface protein charges or charge transfer through the electron donor/acceptor groups of the proteins that can exert gating effects. Protein adsorption can also affect the dielectric constant of the electrical double layer in aqueous solution thereby changing the gate efficiency of the electrolyte or introducing scattering centres that can decrease the device conductance. Moreover, adsorption of species at the metal contact/nanotube interface can induce conductance modulation. Adsorbed species can change the metal work function of the contacts and alter the Schottky barriers at the metal–nanotube interface leading to a significant change in the conductance of the devices. However, passivation of metal contacts inhibits signals due to metal work function modulation and the Schottky barrier effect can be suppressed.

The FET device was incubated with different concentrations of Mb in 0.01 M buffer phosphate. The liquid-gate transfer characteristics were acquired continually during the incubation up to reach a stable response which took about 15 min. Figure 7.6 shows the transfer characteristics in presence of only buffer and after being incubated at different concentrations of Mb. The I_{SD}/V_G curves for each incubated concentration were recorded after signal stabilization (that is after the 15 min of incubation) at a variable liquid-gate voltage and with a fixed source/drain voltage (V_{SD}) of 0.1 V. The starting CNT-FET device in PBS presents

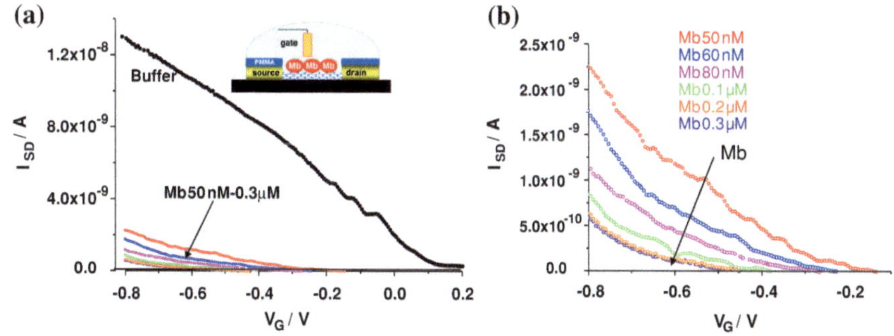

Fig. 7.6 **a** CNT-FET response in presence of only buffer and after increasing incubation concentrations of Mb. **b** A zoom of plot (**a**)

p-type conduction. The interaction of the CNT with increasing concentrations of Mb produces a shift in the voltage gate and also a change in the conductance slope. Such modulation in the conductance can be the result of different contributions that sometimes become difficult to individually distinguish. As mentioned in Chap. 1, for an unambiguously assignment of the operating mechanism one needs to have a CNT-FET bipolar response which is not our case. The measurements are taken at a pH which is close to the protein electroneutrality. Therefore, a screening effect in the nanotube because of the surface charges on the protein is negligible. However, the protein contains amine groups that can donate electrons to the CNT and dope negatively the CNT. This charge donation can be detected as a shift in the threshold voltage of nanotube transistors toward more negative gate voltages. The decrease in conductance of the CNT upon Mb attachment cannot simply be an effect of electrostatic gating by Mb. We also propose that the conductance decrease can be a result of a change in the capacitance of the tube that occurs when Mb is immobilized at high extent, which is quite probable at the concentrations used for the incubation. Mb on the surface of the CNT inhibits ions in the liquid to come close to the CNT, thereby decreasing the double layer capacitance. As a result, the total capacitance of the system, C_t, decreases and so does the conductance of the CNT. We can not rule out the development of scattering centres as consequence of the interaction of Mb with the tube since a change in the conductance slope can also be observed in the figure.

We have also taken advantage of the sensitivity of the CNT-FET device in order to probe possible conformational changes of the adsorbed protein with pH. In previous results we have shown the electrochemical behaviour of Mb on CNT platforms with respect to oxygen electrocatalysis as a function of pH. The results showed that for pH values below 6.0, where the protein was positively charged, the protein underwent conformational changes which decreased its affinity for oxygen and the electrocatalysis of oxygen was suppressed. This behaviour has been depicted in the inset of Fig. 7.7 where the electrochemical behaviour for pH = 7.3 and 5.4 is observed. We have also followed up such changes by

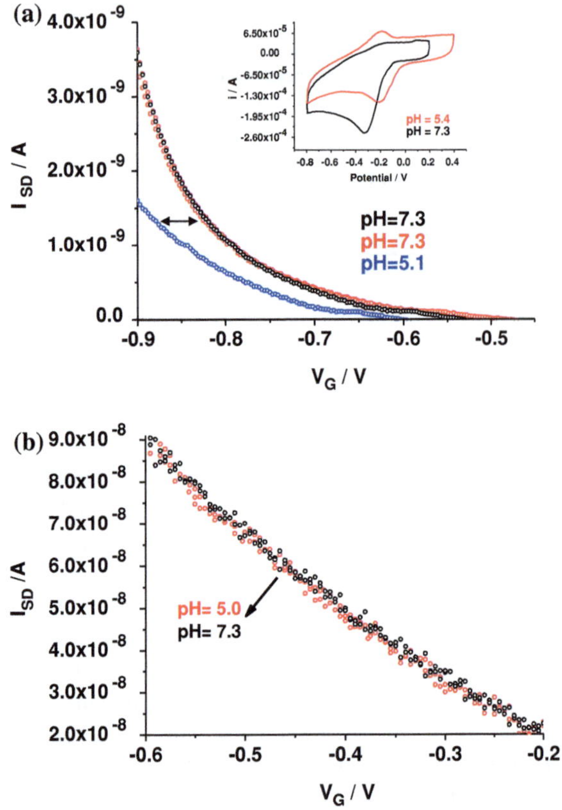

Fig. 7.7 **a** Reversible CNT-FET response of Mb adsorbed on CNT when changing pH. *Inset* Cyclic voltammetry of Mb immobilized on a CNT electrode. **b** CNT-FET response in absence of Mb at the two different pHs

monitoring the electronic response of CNT-FET which has been incubated in 0.3 μM of Mb at the pHs of 7.3 and 5.1. In Fig. 7.7a it can be observed that the system reversibly responds to those changes of pHs. The conductance change can be attributed in principle to conformational changes in the protein probably induced by the change of the net protein charge. The shift in the gate voltage is the one expected for positively charged species on the CNT. No appreciable changes were observed when the CNT-FET response was monitored in absence of the protein at those pHs.

7.3 Label-Free Electrical Detection of Proteins: FET-Aptasensors

Single-walled carbon nanotube field-effect transistors have started to be used to monitor aptamer–protein affinity-binding processes [1–3]. Protein limits of detection in a range of 50 to 0.25 nM have been reported in such papers which encourage the use of such devices with improved designs. As mentioned before, the reason of using aptamers in FET-type sensing assay formats lies in their small size which becomes a crucial topic when trying to detect interactions between biomolecules with a liquid FET sensor. As mentioned in Chap. 1, the FET sensor is sensitive to processes occurring in the double layer; therefore it is important that the different species that undergo a biorecognition process fit the dimensions of the double layer thickness. We discussed previously that in order to fit such processes in the double layer dimensions we should work with small receptors for the biosensing process or decrease the ionic strength to increase the double layer dimensions. In the last example, in which we monitored the CNT-FET response to Mb, we could use a buffer with considerable ionic strength because the protein was small and it was in close contact with the CNT walls. However, if one has to monitor a biointeraction which is not in close contact with the FET, the sensor sensitivity can dramatically decrease if we do not adjust the ionic concentration in the solution. That is why it is difficult to monitor protein–protein binding processes in immunological FETs (Immuno FETs). The size of antibodies ($\approx$10 nm) is much larger than the electrical double layer so that most of the protein charges will be at a distance greater than the Debye length ($\approx$1.8 nm for 1$\times$ PBS). The recognition might occur outside the double layer resulting in small potential changes on the gate. However, the dimensions of aptamers (~2 nm with their folded structure) enable the recognition binding events occur within the Debye length of the double layer.

Accordingly, we have used a CNT-FET modified with aptamer receptors for following up protein recognition. The FET device, the passivation protocol and the modus operandi to monitor the in situ bio-recognition process were the same than the one just discussed in the section above. However, we have chosen a diluted buffer solution of 0.05$\times$ PBS in order to achieve lower ionic strength and therefore increase the Debye length. Under these conditions, the Debye length was about 3.5 nm[4] (just for comparison at 1$\times$ PBS the Debye length is about 1.8 nm). Thicker Debye lengths could be achieved with more diluted PBS but a compromise needs to be reached since the low concentration of ions can decrease the efficiency in the biorecognition process.

Before going in the details of the CNT-FET response, we will recall the functionalization process of the device. The derivatization protocol was carried out by using a linker based on pyrene with a carboxylic functional group as illustrated in Fig. 7.8. The CNT-FET is dipped on an aqueous solution of pyrenebutyric acid and thoroughly rinsed in water. At this stage, the carboxylated pyrene is supposed to be adsorbed[5] on carbon nanotubes sidewall and can be used to covalently anchor $-NH_2$ terminated aptamers by carbodiimide chemistry.[6] After that PEG

[4] At room temperature and in aqueous solutions the Debye length (λ_D) can be expressed as $\lambda_D(\mathrm{nm}) = \frac{0.304}{\sqrt{I(M)}}$ where λ_D is in nanometers and the ionic strength (I) in mol/L.

[5] We have shown in the experimental section that pyrene adsorbs on the CNT walls, a process that has been checked with confocal fluorescence microscopy.

[6] Just for recalling some experimental tests, the carbodiimide coupling was successfully followed by XPS measurements.

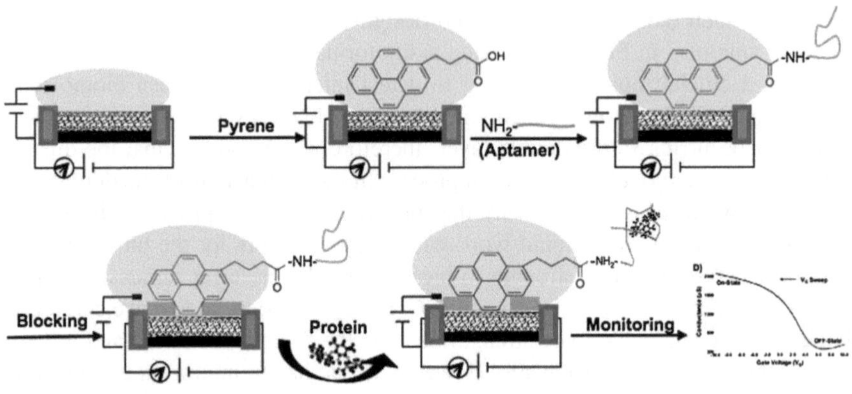

Fig. 7.8 An scheme showing the biofunctionalization steps of the CNT-FET for monitoring aptamer/protein interactions

Fig. 7.9 Open circuit potential for a gold electrode versus Ag/AgCl reference electrode at a thrombin concentration of 1×10^{-8} M

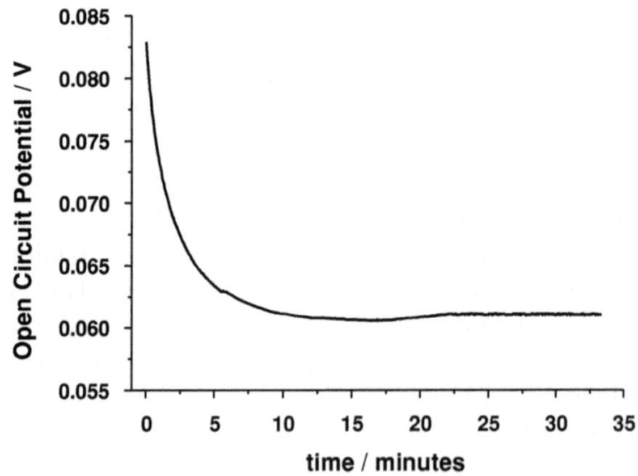

polymer is non-covalently immobilized onto the CNT. That polymer becomes a quite effective barrier against non-specific protein adsorption by forming a highly hydrophilic (but charge neutral) and sterically hindering layer at the surface. After such stage, the functionalized CNT-FET device is ready to be exposed to the target protein for monitoring protein–aptamer interaction.

We have additionally tested the role of the gold electrode as pseudo-reference electrode. For that we have measured the open circuit voltage of the gold electrode vs. an Ag/AgCl reference electrode in a buffer solution containing the target protein. We have observed a short transient of some seconds and then the signal became stable as shown in Fig. 7.9. Accordingly, we can consider that the in-chip gold electrode can be used as a pseudo-reference electrode for liquid gating. The potential almost does not change at the time at which the measurement is taken place (around 15–20 min).

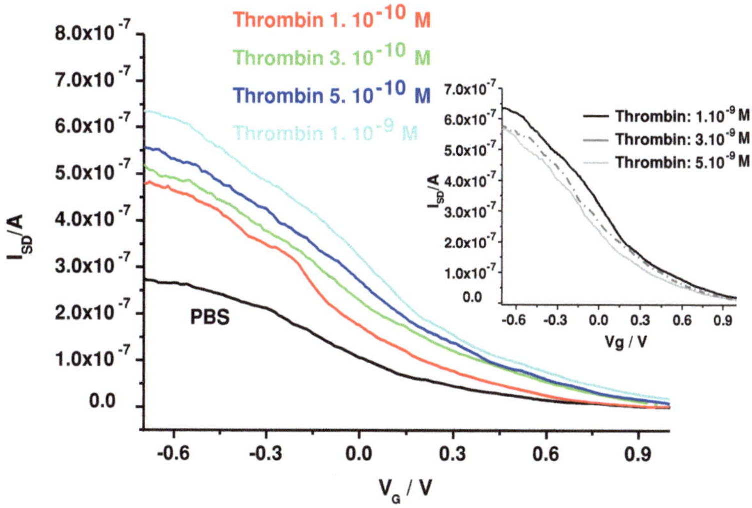

Fig. 7.10 Changes of conductance of a CNT-FET modified with thrombin aptamer as a function of different concentrations of thrombin. The *inset* shows the changes of the conductance at higher thrombin concentrations

Figure 7.10 shows the change of conductance of the FET when the CNT is functionalized with a thrombin aptamer and measured in presence of PBS only (black curve) and in presence of increasing concentrations of thrombin. The procedure was very similar to that reported with myoglobin. The aptamer functionalized FET was incubated with different concentrations of thrombin in 0.05 × PBS and the I_{SD}/V_G profiles were recorded continually during the incubation up to get a stable response. The figure shows the conductance profiles recorded at the stable conditions as a function of different thrombin concentrations. It can be observed that the conductance is shifted to more positive gate voltages with increasing concentration of thrombin in the low concentration range. At higher target protein concentrations, the I_{SD}/V_G profiles start to be shifted towards more negative gate voltages which denotes a change in the sensing mechanism (see inset of Fig. 7.10). A shift of the conductance profiles towards more positive voltages can be a consequence of more negatively charged species coming closer to the nanotube in the double layer region. Such charges can generate a screening effect on the nanotube doping it positively. The increase of the p-doping in the conductance signal with the increasing thrombin can be explained by considering two effects. On one side, the starting immobilized aptamer containing 18 bases can be quite flexible in the aqueous environment. The thrombin aptamer can fluctuate between two states, an extended (random coil) and a tertiary quadruplex structure kept by hydrogen bonds (as explained in previous sections). Such folded structure is favoured by some cations in the buffer solution. However, as soon as thrombin is added in the medium, the equilibrium is shifted more toward the quadruplex form. It has been recently demonstrated that apart from some cations causing such aptamer folding

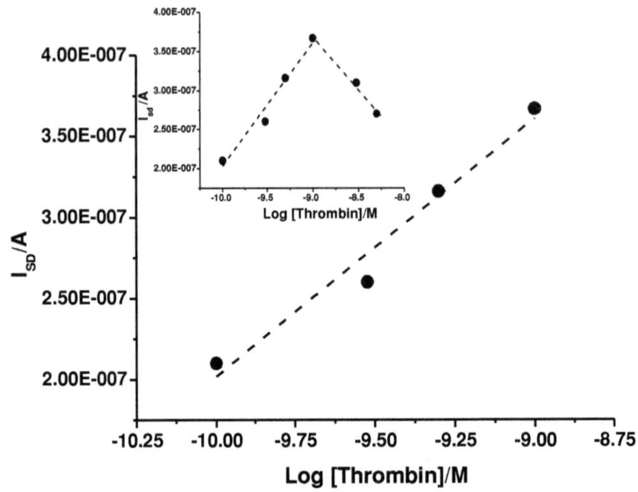

Fig. 7.11 Calibration curve showing the change of I_{SD} as a function of thrombin concentration. The calibration was obtained at a gate voltage of −0.05 V. The *inset* shows the change of the sensing mechanism at higher protein concentrations

(which is also favourable for the protein binding), the thrombin itself also assists or induces such quadruplex structure to some extent [4]. With such folding, the aptamer is bent in a more compact form in the double layer with the consequent increase of charges in close proximity of the nanotube. On the other hand, such increase of the negative charges in the Debye length can come from the protein itself. Thrombin exhibits an isoelectric point of 7–7.6. In 1 × PBS the pH is around 7.4. However, in the diluted 0.05 × PBS the pH increases up to 7.8,[7] conferring to the protein a slightly negative charge [5]. Consequently, as the biorecognition event is taken place, more negative charge is injected in the double layer producing such p-doping effect in the CNT charge carriers.

However, the possible effects described above can only be dominant for a certain range of protein concentrations. After exposing the CNT device to higher concentrations, the presence of the interacting protein in the double layer can slightly change the capacitance of the double layer lowering the conductance of functionalized CNT device as can be observed in the inset of Fig. 7.10.

We have taken the lower range of concentration in order to get the calibration curve shown in Fig. 7.11. The standard deviations of the conductance curves were in the range of 10–15 %. A limit of detection of 20 pM was calculated by

[7] When the concentration of a buffer varies, the ratio of the concentration of buffer components remains constant. However it is the variation in the ratio of the activity coefficient of the solution components that leads to variations in the pH value of the buffer solution. The effect is an increase of the pH depending on the dilution factor.

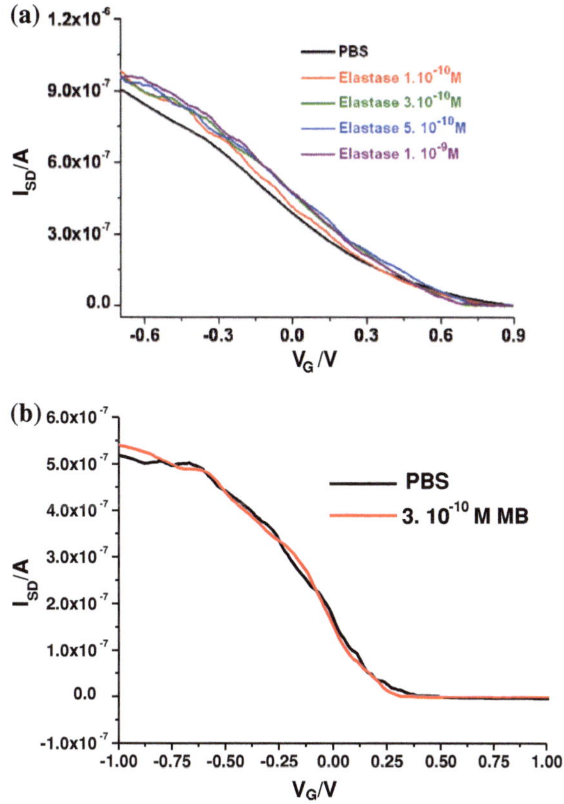

Fig. 7.12 Response of the CNT-FET sensor modified with thrombin aptamer to two different proteins such as elastase (*top*) and myoglobin (*bottom*)

considering the expression $3\sigma_{blank}/S$, where σ_{blank} is the standard deviation in blank solution and S is the sensitivity calculated from the slope of the sensor response (I_{SD} vs. [Thrombin]). The inset of the Fig. 7.11 also shows the I_{SD} profiles for higher thrombin concentrations at which the sensing mechanism changes.

To measure the selectivity of thrombin aptamers, we used different target proteins, such as myoglobin and elastase. Elastase is another serine protease with an isoelectric point and molecular weight similar to that of thrombin.

Figure 7.12 shows the CNT-FET response to non-specific protein targets. The results indicated that negligible non-specific binding interactions are taken place and one can consider that the sensor response is highly specific to the binding of thrombin protein.

In this kind of systems the non-specific adsorption remains minimized probably aided by the low porosity degree of the CNT network as compared to the ones used in the CNT microelectrodes or CNT modified glassy carbon electrodes.

In summary, the experimental results show that the optimized CNT-FET layout together with the immobilization scheme of the receptors and electrolyte conditions result in a very good performance of the sensor for detecting the biorecognition event between the thrombin aptamer and its specific thrombin protein. Such good performance is supported by the low limits of detection that were achieved (20 pM) and their high selectivity when using other target proteins.

7.4 General Conclusions of the Electric Response of Carbon Nanotube Field-Effect Transistors

In this section, we have introduced the use of passivated devices for following up in real-time a biorecognition process. Such design can decrease, in a big extent, side electrochemical signals that can smear the true signal of the sensor, a topic that has not so much been considered in previous studies of CNT-FET aptasensors. On the other hand, we have paid special attention to the critical feature of the Debye length. As mentioned, the Debye screening constitutes an important design consideration for label-free detection. For maximizing the detection signal is very important to optimize buffer concentrations, linker and molecular arrangements. All these considerations have allowed us to follow the sensor signal with more sensitivity, capturing changes in the sensing mechanisms and achieving lower limits of detection as compared with previous studies based on aptamers CNT-FETs.

References

1. Maehashi K, Katsura T, Kerman K, Takamura Y, Matsumoto K, Tamiya E (2006) Anal Chem 79:782–787
2. So H-M, Won K, Kim YH, Kim B-K, Ryu BH, Na PS, Kim H, Lee J-O (2005) J Am Chem Soc 127:11906–11907
3. Yoon H, Kim J-H, Lee N, Kim B-G, Jang J (2008) ChemBioChem 9:634–641
4. Chi C-W, Lao Y-H, Li Y-S, Chen L-C (2011) Biosens Bioelectron 26:3346–3352
5. Bates RG (1954) Anal Chem 26:871–874

Chapter 8
Conclusions

8.1 General Conclusions

The advent of carbon nanotubes has opened innovative perspectives for research and development of carbon electrodes. In this context, this thesis has attempted to highlight the electrochemical properties of carbon nanotubes for biosensing purposes by rooting them mainly on their structural, electronic and chemical properties.

Due to their structure, CNTs show remarkable surface characteristics. For instance, they exhibit chemical and electrochemical anisotropy with enhanced reactivity at the edges as compared to walls. Therefore, a high density of CNT edges is expected to increase the electrochemical performance of the electrodes providing more electrochemical sites for increased electron transfer kinetics. Such effect turns out to be similar to the behaviour found on edge HOPG. Moreover, oxygenated surface functionalities can increase even further the electrochemical performance depending on the redox system.

However, the nanometre size and high aspect ratio of the carbon nanotubes are the distinct features which have mostly contributed for innovative electrochemical applications and for establishing the differences with respect to other carbon materials. Such characteristics make CNTs exhibit a very large collective surface area for covalent or non-covalent (bio)molecule functionalization and for electrochemical reaction amplification. Moreover, CNTs represent a versatile material which can be tailored in different geometrical electrode arrangements for exploiting their sites with higher electroactivity. All these ingredients make CNT systems ideal candidates for amplified signal transduction in the electrochemical/biochemical sensing field.

Accordingly, we have focused on the electrochemical performance of different CNT electrode systems as transducers and on their suitability as enzymatic/redox protein biosensors, DNA-detection sensors and aptamer-based sensors.

M. Pacios Pujadó, *Carbon Nanotubes as Platforms for Biosensors with Electrochemical and Electronic Transduction*, Springer Theses, DOI: 10.1007/978-3-642-31421-6_8,

8.2 Specific Conclusions

8.2.1 To Study the Use of Carbon Nanotubes as Electrochemical Transducers and the Relation Between Their Structure and Their Electrochemical Reactivity

We have compared the performance of composites made with the same weight ratio of CNTs versus epoxy resin, using MWNTs of different external diameter and structure (concentric and bamboo-like ones) and SWNTs, all of them with similar lengths. The CNTs were strongly purified to minimize the possible electrochemical contributions of the metal catalyst impurities.

Our results indicate that, beyond the obvious beneficial role of high surface areas for amplifying the electrochemical response, the presence of a high density of CNT edges is very important for enhancing the electrochemical transducer characteristics of these composites.

Specifically, a clear trend was found in these composites: the smaller the CNT diameter (i.e. higher density of CNT in the composite and hence more density of edge sites) the higher the rate transfer constant, the higher the peak current and the smaller charge resistance. In fact, the bamboo-like CNT composites were the ones showing the fastest electron transfer rate and the smallest charge resistance due to their intrinsically high density of graphitic edges. The response of the CNT composites was compared with HOPG (basal and edge) and similar composites made from graphite powder. The electrochemical performance of the CNT composites is better than that of the graphite composite and lies between the basal and edge HOPG but closer to the edge HOPG in response.

All these results settle some important requirements and basis to build efficient electrochemical transducers based on carbon nanotubes and indeed suggest that the structure and orientation of the carbon nanotubes can play a very important role that should be taken into account. That was the reason for extending our studies to the use of vertically aligned CNT systems either in a macroscopic or microelectrode array formats whose conclusions will be pointed out later on. Moreover, these results also encouraged us to combine the electrochemical properties of CNTs with the enhanced electroanalytical properties of ultramicroelectrode arrays (fast mass transport, fast response and high signal/noise ratio). Accordingly, we have established the first foundations in the development of vertically-oriented CNT ultramicroelectrode arrays for ultrasensitive detection, fabricated with inexpensive and more accessible methodologies. Our preliminary results have demonstrated that the opal inversion technique based on the self-assembly of polystyrene beads as moulds can provide large area periodic arrays of vertically aligned CNTs grown by plasma enhanced chemical vapour deposition. We have also developed a passivated protocol to cover the walls of the CNTs and part of the voids between each nanoelectrode. The electrochemical response of the CNT-UMAs showed a sigmoidal CV profile which is related with a steady-state current dominated by radial diffusion, the typical feature of UMAs electrodes.

Such passivated and upright CNT nanoelectrode arrays with the more electrochemical reactive sites facing towards the analytes hold promise for improving the sensitivity and detection limits up to the range of few molecules. Additionally they can act as molecular wires where electroactive biomolecules can be plugged at their ends facilitating the direct electron transfer. Moreover, these UMAs can exhibit the additional feature that the porosity degree or surface area effects can be better controlled and thus improving, in some cases, selectivity in biorecognition processes.

8.2.2 To Study the Use of Carbon Nanotubes as a Suitable Platform for Detecting Biorecognition Events and the Use of Electrochemistry as a Signature of Carbon Nanotube Functionalization and Sensing

We have tried to highlight the electroanalytical CNT properties for biosensing applications. As mentioned earlier, the remarkable electrochemical properties of CNT arise as a combination of different aspects given mainly by their surface structure, nanometric size and aspect ratio, their collective high specific area, their electrochemical anisotropy, their versatility to be disposed in different electrode configurations and their robustness for (bio)chemical functionalization.

We have illustrated three examples that demonstrate the suitability of CNT as electrochemical transducers in enzymatic sensors, DNA-sensors and aptasensors.

We have used physisorption and covalent attachment through carbodiimide chemistry as biomolecule immobilization strategies. Amide formation has been successfully used for the anchoring of proteins, amine terminated oligonucleotides, etc., with the CNT carboxyl groups.

In the first example, we have proven that carbon nanotubes-based materials exhibit an enhanced electrochemical response towards metalloproteins (myoglobin and catalase) as compared to other electrode substrates based on the analysis of the electron transfer rate constant. The CNT systems represent an adequate platform for the protein wiring which allows on one side the direct electrochemistry with the hemin active redox center of the protein and on the other side the preservation of the protein bioactivity to different analytes (oxygen, hydrogen peroxide, etc.). We have tested two different CNT electrodes: forest CNT and epoxy CNT composite electrodes and we have also established comparisons with respect to graphite epoxy composites. In both CNT systems, we have observed enhanced electrochemical response. In the forest CNT electrode, the enhanced electrochemical signal could arise from a more facilitated electron transfer. This seems to be a consequence of the exposition of a higher density of the more electroactive sites (edge sites), just at the electrode/solution interface. That could facilitate the wiring and electron exchange with metalloprotein.

In the case of the CNT composites, the electrochemical improvement comes from the more porous character of this material and consequently higher surface area for the immobilization of more amount of protein which amplifies the signal response. All these aspects exert influence in the final electroanalytical characteristics such as the remarkable low detection limits in the response of the protein to its specific substrates (case of the metalloprotein/forest CNT electrodes), the high sensitivity due to the large protein loading (case of the metalloprotetin/CNT composite electrodes) and the wide analyte concentration range.

Metalloprotein/CNT electrodes also represent an appropriate system to monitor the oxygen sensitivity of protein as a function of the pH. Conformational changes in Mb with the consequent loss of its affinity to oxygen were clearly followed at low pH. The effect of consecutive scan cycles as well as cathodic or anodic polarization on Mb/CNT electrodes gave us insights of the time scale of oxygen diffusion in the electrolyte diffusion layer and its uptake by Mb.

We have mentioned that CNTs exhibit high versatility for developing different CNT electrode configurations. Thus, the second example reported here is about the use of electrochemical CNT transducers disposed in a microelectrode array for DNA detection. That electrode system has resulted in a robust platform for single point linkage of probe DNA and for following up the hybridization process with a simple, clean and reversible detection scheme based on electrostatic groove redox binders which are positively charged. Different electrochemical techniques have been used such as Cyclic Voltammetry, Chronocoulometry and Impedance Spectroscopy. Chronocoulometry has allowed us to discriminate the accumulated redox species at the interface from those diffusing in the electrolyte. The monitoring of the accumulated species has provided us a sensitive way to follow the DNA hybridization. On the other side, the combination of CNT microelectrode arrays and impedance measurements has turned out to be a powerful approach to follow the hybridization process with such highly positively charged redox indicators. It is worthy to mention that without such CNT microelectrode and its scaling down size effects on the impedance measurements, it would have not been possible to characterize the biorecognition process with that simple but sensitive approach. Again, all these results encourage the use of miniaturized systems and hold promise for the development of simple and practical genosensors thus avoiding effortful and expensive DNA labelling schemes or DNA damage by oxidation.

It is often necessary to directly detect proteins instead of DNA in order to fully understand cellular function, to detect diseases, or to monitor their progress or to detect a drug's effect in an organism. The third example is focused on the detection of proteins using aptamers, as an alternative of antibody-based strategies.

In this case, CNT dispersed on glassy carbon electrode and microelectrode arrays made of vertically aligned CNTs have been used as electrode platforms. Impedance spectroscopy was successfully used to detect the interaction of aptamers with their specific proteins such as Thrombin and Lysozyme. The systems exhibit remarkable sensitivity at low protein concentration range, though some interaction with non-specific proteins could not be completely eliminated.

We have observed that the aptasensor impedimetric response is very dependent on the previous treatment of the aptamer, the blocking agent for non-specific adsorption, the ionic strength and pH of the buffer solution. We have presented here our first preliminary efforts for tuning all these aspects to achieve appropriate sensor response. However there are still open issues that need further optimization for a better sensor performance. One of them is the non-specific adsorption minimization which can be improved by testing other blocking agents or by addressing CNT platforms with lower degree of porosity. A lower degree of porosity could even decrease the limits of detection. In spite of the present limitations both CNT sensor platforms have provided promising detection limits. Perhaps, the microelectrode CNT configuration represents a more attractive candidate since impedance signal is amplified with miniaturization and at the same time is a more tunable system for controlling porosity or for generating CNT ultramicroelectrode arrays with better analytical performance.

8.2.3 To Optimize Carbon Nanotube-Field Effect Transistor (CNT-FET) Layout for Following up in Real Time a Biorecognition Process

Liquid-gated CNT-FETs have been optimized for their use in aqueous environment by designing a passivation protocol based on the deposition of a PMMA layer on the whole sensor except for the channel of the CNT-FET, for part of the liquid polarization electrode and for the other big contact pads, source and drain. Such design minimizes side electrochemical signals that can smear the true signal of the sensor and allow the real time detection of biorecognition processes. The liquid gate can become over two orders of magnitude more efficient than the back gate, used in dry conditions; this means an enhanced gate coupling between the liquid and the CNT and higher transconductance characteristics which are crucial factors for determining the ultimate sensitivity of the sensor.

As applications, the adsorption of a protein (Mb) on the CNT walls has been successfully monitored in real time. Moreover, the performance of the CNT-FET as aptasensor has been evaluated by monitoring the biorecognition event between an aptamer and its specific target protein.

On one hand, it has been proved that the interaction of the CNT with increasing concentration of Mb produces a shift in the voltage gate and also a change in the conductance slope that could be used for following the protein adsorption. Such modulation in the conductance can be the result of different contributions that sometimes become difficult to individually distinguish (electrostatic gating effect, capacitance mechanism or mobility mechanism). We have also taken advantage of the sensitivity of the CNT-FET device in order to probe possible conformational changes of the adsorbed protein with pH.

On the other hand, CNT-FET has been also successful in detecting the biorecognition event between the thrombin aptamer and its specific thrombin protein. However, special attention to the critical feature of the Debye length had to be paid. Debye length has to be thick enough to allow the label-free recognition event to take place within the double layer. For maximizing the detection signal is very important to optimize buffer concentrations, linker and molecular arrangements. Thicker Debye lengths have been achieved with more diluted PBS but a compromise needs to be reached since the low concentration of ions can decrease the efficiency in the biorecognition process. The derivatization protocol has been carried out by using a linker based on pyrene with a carboxylic functional group. All these considerations have allowed us to follow the sensor signal with more sensitivity, capturing changes in the sensing mechanisms and achieving lower detection limits as compared with previous studies based on aptamer CNT-FETs.

Chapter 9
Perspectives

9.1 Perspectives in Fundamental Electrochemical Issues

In this thesis, we have tried to highlight the electroanalytical capabilities of CNT electrode platforms for biosensing applications. Behind these promising electrochemical devices, there are still many fundamental issues that affect the electrochemical response and which are not so easy to be controlled. Such issues can be found right from the first steps of CNT growth, purification, separation, etc. Given the interfacial nature of electrochemistry we have made an effort to correlate CNT structure and surface with the electrochemical reactivity by establishing general trends. However, the scope of this thesis was not framed in merely fundamental electrochemical issues, and a more exhaustive study is needed to settle these points. Despite the fact that CNT edges and defects play a very important role in the electrochemical response, we are also aware that metal catalyst impurities (that cannot be completely removed mainly due to their sheathed nature) can also have some contribution. We have tried to minimize such effect, but it would be very interesting to go deeper in the fundamental studies and try to distinguish more precisely the electrochemical contributions of edges, defects, surface oxides, walls and sheathed catalyst particles. Special efforts should be taken to clarify the potential electrocatalytical mechanisms of carbon coated metal impurities to certain redox species. Such identification is very challenging and will require the use of experimental techniques that can visualize electrochemical activity at a local scale and relate this to the underlying structural and electrical properties of CNTs. Fundamental electrochemical studies on CNTs could be facilitated if CNT growth and processing could be better controlled. Advances in such issues could further improve nanotube uniformity, decrease the dispersion in the results of different research groups, improve the device efficiency and also provide further insights in electrochemical fundamental questions.

M. Pacios Pujadó, *Carbon Nanotubes as Platforms for Biosensors with Electrochemical and Electronic Transduction*, Springer Theses, DOI: 10.1007/978-3-642-31421-6_9,

9.2 Perspectives in the Electrode Configurations

We have shown the versatility of CNTs to be disposed in different electrode configurations: from a vertical disposition and high density of CNTs (i.e. forest CNT), CNT composites, ultramicroelectrode arrays up to a field effect transistor configuration. Compared to the other CNT electrochemical systems, the use of vertically aligned CNT nanoelectrode arrays are of particular interest since they can increase the detection to the range of few molecules. Moreover, they permit to anchor biomolecules with a high degree of stability, flexibility and control of space distribution. Additionally, the extent of porosity and the development of better defined non-specific adsorption blocking layers can be also more controllable. That can help to minimise non-specific interactions and increase selectivity, especially when using proteins as target species. We have provided the first grounds in the construction of CNT UMAs using more accessible fabrication approaches but, as we mentioned, it is an open research line which needs to be improved and exploited given the promising characteristics of these systems in bioanalytical assays. In this context, one could take advantage of the different sensing strategies that have been exploited with self-assembled monolayers (SAMs) on gold electrodes and more specifically with mixed SAMs (composed of non-conducting alkanethiols and of more conducting ones such as conjugated thiols). The passivated CNT-UMA system could exhibit a thin layer format similar to the one formed by those mixed SAMs but with the advantage of higher conducting and more robust paths. Moreover, these platforms could profit from a higher control of their conductive spot spacing and also of the density of the immobilized biorecognition elements. Therefore, it can be envisioned that in the near future, these UMAs could represent the basis for the design of (bio)sensors with high control and highly improved sensitivity.

CNT-FET devices exhibit promising perspectives as label-free biosensors and constitute an interesting alternative to the typical electrochemical architectures which normally require electrochemical tags to transduce the biorecognition event. However, there is still much pending work in understanding at a microscopic level the mechanisms behind the sensing signal. Improving our understanding of the electrostatics interactions of carbon nanosensors with aqueous environments not only allows the study of fundamental charge interactions at the nanoscale, but it also aids in setting up and interpreting biosensing experiments, studying sensor specificity and optimizing sensor layout to increase sensitivity and to lower detection limits.

However, we are still in the starting point of the development of these devices. They constitute an open topic which needs to be optimized. The sensitivity of the FET can be improved by changing its geometry, interelectrode gap, the size of free-PMMA openings in the CNT channel or by changing the blocking agents, etc. The devices can be also engineered in such a way that the very sensitive mechanism based on Schottky barrier modulation is not totally suppressed. For that, one can leave a very tiny fraction of the metal contact/CNT region without passivation but checking that the side electrochemical reactions do not overpass the limits that mask the sensor response. Moreover, our CNT-FET device has all the base architecture for the development of an integrated, high throughput, multiplexed, real time biosensor in combination with microfluidics. As the development and optimization studies of our CNT-FET nanosensor proceed, we predict that sensitive multiplexed detection of numerous clinically important biomolecules would become possible in a rapid and high throughput format. At this stage, we can only say that this issue is just at the beginning of its potentialities and leaves an open avenue for future new studies.

9.3 Perspectives in the Biorecognition Elements Discussed in this Thesis

We have used the forest CNT electrodes and epoxy CNT composites as a protein/enzymatic (myoglobin, catalase) sensors taking advantage of the facilitated biomolecule wiring with the vertical disposition of the CNTs and the porous character of the composites. The enhanced electrochemical performance of these heme protein/carbon nanotube electrodes hold promises for development of efficient oxygen and hydrogen peroxide sensing devices that could exploit the electrochemical advantages inherent to these systems in biomedical and environmental applications. These advantages could even be extended to other sensing analytes of biomedical importance such as free radicals or other highly reactive oxygen species (ROS). Many metalloproteins such as superoxide dismutase, glutathione peroxidase, and the catalase investigated in this work, among others, are involved in the destruction of hydrogen peroxide, generated during cellular metabolism and consequently oxidize different toxins, such as formaldehyde, formic acid, phenols, and alcohols. That is why metalloproteins or heme proteins are very important ROS scavengers and can be used at the same time as sensors of these reactive species. We have also proved that other heme-based proteins such as myoglobin with its redox activity are also capable to have similar functions to the mentioned antioxidant enzymes. Nitric oxide (NO) is another candidate to be sensed with this kind of heme protein/CNT devices. Accordingly, in order to clarify the function and to control the concentration in biological systems of NO and ROS species, selective and sensitive detection is of great importance. There exist some free radical diagnostic techniques such as electron spin resonance, chemiluminescence, chemical labeling by quenching with free radicals followed by spectroscopic methods, using trapping agents, etc. However, electrochemical biosensors based on protein-CNT electrodes have been proposed in this thesis because their inherent sensitivity, low cost, easy handling and the possibility of giving results in real time. Indeed, the synergy of the electrochemical techniques with novel biomolecule/carbon nanotube platforms constitutes key promising sensing strategies with high impact in biomedical applications.

On the other side, we have explored other sensing strategies with CNT transducers disposed in a microelectrode array for DNA-detection and aptamer-based sensors. These systems take advantage of the high sensitive electrochemical impedance spectroscopy technique for label-free and non-destructive detection of oligonucleotides and proteins. The determination of nucleic acid sequences from humans, animals, bacteria and viruses constitute the starting point to solve different problems such as food and water contamination caused by microorganisms, detection of generic disorders, tissue matching, forensic applications, etc. The detection of DNA hybridization is of central importance to the diagnosis and treatment of genetic diseases, for the detection of infectious agents, and for reliable forensic analysis. Efficient and inexpensive devices for determining RNA and DNA sequences are required for a better decentralized diagnosing, preventing and treating of many human diseases. Such devices can be used in other applications such as veterinary and forensic medicine, environmental testing, etc. In this context, we expect that miniaturized CNT electrodes (microelectrodes, UMAs, FETs) can become promising detection schemes to evaluate the ultimate capabilities for base mismatch discrimination.

While aptamer-based biosensors emerged only about 10 years ago, they have already found broad applications in basic research, environmental monitoring and biomedical diagnostics. In spite of these rapid advances, aptamer-based bioassays are still immature when compared to immunoassays, which in a sense reflects the limited availability of aptamer types and the relatively poor knowledge of surface-immobilization technologies for aptamers. Nanotechnology has made an important addition to the analytical and diagnostics workbox (e.g., the combination of aptamers with novel nanomaterials has led to highly sensitive and selective aptasensors). Therefore, CNTs are showing to be good candidates to be used as transducers for aptasensors in the more conventional electrochemical detection and also with many advantages in the CNT-FET device configuration. Here, we have presented our preliminary efforts to achieve protein detection with aptasensors. Still there is much pending work trying to control selectivity, a topic that is not so well controlled in massive CNT electrode configurations where the extent of porosity can exhibit, in such case, some drawbacks.

Under this context, another interesting approach might be the use of the electronic tongue concept, in which an array of aptasensors could be used to determine simultaneously the target proteins plus their main interferents. Moreover, it is of large importance to continue evaluating fundamental issues concerning the DNA binding conformation, the effect of ions and pH to have better clues of the interaction mechanism between protein-DNA and thus increase the sensor device efficiency.

In summary, we have made a general outlook of the different topics covered in this thesis work. Indeed, all the systems here discussed encourage the exploitation of novel and miniaturized CNT electrode configurations that can be tunable according to different (bio)electroanalytical applications. However, we would like to stress that such applications do not remain restricted to the typical electrochemical biosensing field but can cover other exciting areas such as the development of CNT tools for brain stimulation and neuronal signal detection or the development of novel electrochemical actuators or electrochemical energy-harvesting devices, topics of great relevance in today's research.

Index

M. Pacios Pujadó, *Carbon Nanotubes as Platforms for Biosensors with Electrochemical and Electronic Transduction*, Springer Theses, DOI: 10.1007/978-3-642-31421-6,

MIX
Papier aus verantwortungsvollen Quellen
Paper from responsible sources
FSC® C105338

If you have any concerns about our products,
you can contact us on
ProductSafety@springernature.com

In case Publisher is established outside the EU,
the EU authorized representative is:
Springer Nature Customer Service Center GmbH
Europaplatz 3, 69115 Heidelberg, Germany

Printed by Libri Plureos GmbH
in Hamburg, Germany